CHICAGO with KIDS

places to go and things to do

SHERIBEL ROTHENBERG AND ELLEN DICK

Chicago Review Press

Library of Congress Cataloging-in-Publication Data

Rothenberg, Sheribel.
Chicago with kids : places to go and things to do / Sheribel Rothenberg and Ellen Dick. — 1st ed.
p. cm.
ISBN 1-55652-057-3 : $8.95
1. Chicago (Ill.)—Description—Guide-books. 2. Family recreation—Illinois—Chicago—Guide-books. I. Dick, Ellen. II. Title.
F548.18.R67 1989 89-33675
917.73'110443—dc20 CIP

Printed in the United States of America
First Edition
1 2 3 4 5 6 7 8 9 10
Published by Chicago Review Press, Incorporated
814 N. Franklin Street, Chicago, Illinois, 60610

ISBN 1-55652-057-3

Cover design and illustrations by Fran Lee

To Eli and Emily:

Joyous companions, critics, and consultants on places to go and things to do.

CONTENTS

4. Chicago's Ethnic Museums 71

1
IN THE BEGINNING . . . IDEAS

PLANNING

The length of a city adventure and its rhythm can be as varied as your mood, your children's temperaments, and the weather. Before starting out, it's a good idea to gauge your crew's needs and endurance.

Every child bent on discovery needs protection from two inevitable side effects of travel: overstimulation and boredom. In order to pace your trip with children, you should keep in mind—and in your children's minds—the following suggestions:

1. Be sure that your children always have something to do, especially for waits in lines or delays.
2. Don't focus on getting home. Expect that your adventure will take one-third again as long with children as without them.
3. Indulge in the unexpected event that may turn out to be the highlight of the outing.

And, while the information in this book is current at the time of publication, we recommend that you always call ahead of time to double-check the hours and the prices.

CLOTHING

As we all know, Chicago's weather can change suddenly, so dress your children in layers and peel them off or add more as the temperature changes.

FOOD

Snacks and drinks are vital to pacing a day's exploration with children. A packed snack of fruit, nuts, raisins, and juice in cans or boxes takes little room in a travel bag and can prevent tears.

CLEANUP

Disposable wet towels have solved the stickiness problems when away from home. Remember to bring them along.

PREPARATION

Read aloud a description of where you are headed to allow the children to prepare for and wonder about the experience to come. Formulate a plan, and if you can, let your children contribute. For example, large public museums are impossible to see all at one time, so let your children help decide which exhibits to see and in what order.

PACING

Just as an exciting story has a beginning, a middle, and an end, so should an outing. The beginning is always getting there—a process in itself if small children are involved. The middle of your outing should be some facet of your destination. The end of your outing can be a relaxing stop for refreshments or lunch and some active play to enable small children to rest comfortably on the way home.

ECLECTIC ADVENTURES

The chapters that follow are loaded with places to go with your children, but sometimes you just want some unusual ideas for things to do. So the next time you and your children have a spare hour or two and are racking your brains for something different to do, consider some of these spur-of-the-moment, eclectic adventures.

If your children are early risers and you can't bear to let them watch television, consider going out to breakfast. Scores of local pancake houses and coffee shops are open in the early hours of the day and provide a bit of local color.

Try, for example, Donna's Notorious Italian Hamburgers at Ashland and Augusta, a six stool, six table restaurant, open 7:00 A.M.–2:00 P.M. daily—their coffee is strong and their bread is freshly baked. Lou Mitchell's, 565 W. Jackson Boulevard, (312) 939-3111, near the Options Exchange, is all a bustle in the early hours of the weekdays and has delicious breakfast options.

Walker Brothers Pancake House on Green Bay Road in Wilmette, next to the Kohl Museum, is a pleasant half-hour drive from the north side of the city and has an excellent and varied breakfast menu, great coffee, and a selection of children's specials. Get there at 7:00 A.M. or be prepared to wait.

The second floor dining room in the Drake Hotel, 140 E. Walton, (312) 787-2200, has windows overlooking the Outer Drive and is very pleasant on a weekday morning. It is not as pricey as you might imagine, considering you can feast on freshly-baked muffins and watch the hapless souls driving to work in the morning traffic.

A relatively new addition to the Chicago breakfast scene is Ace's, 4801 N. Broadway Avenue, (312) 878-8118, at the corner of Broadway and Lawrence Avenues. Ace's is open from 11:00 P.M. until 8:00 A.M., Wednesday through Sunday and uses the slogan, "Where the night owl meets the early bird." To the music of a live jazz trio, you can breakfast on traditional pancakes, bacon, eggs, grits, sausage, and gravy and biscuits, or you can sample catfish and cornbread or seafood and eggs. Perhaps only the adventurous take the wee ones here, but the pleasure of listening to live music in the morning makes the trip worthwhile.

Also consider an early morning visit to one of the great bakeries of the city, such as the Kaufman's Bagel Bakery, 4411 N. Kedzie, (312) 267-1680; the Middle Eastern Bakery—the pita bread factory, 1512 W. Foster, (312) 561-2224; or D'Amato's Bakery, 1124 W. Grand, (312) 733-5456.

Breakfast picnics in the early morning hours of hot summer days at the lake, or even in the neighborhood parks, are a real treat. A walk near a harbor to watch the old fishermen getting their crafts ready for a morning at sea fascinates most youngsters.

Visiting a farmers market or produce market, such as the one along Randolph and Water Streets just west of the loop, is an exciting adventure for a child and a worthwhile trip for a parent.

Some children (and adults) are intrigued by the construction and maintenance of the city. The next time you pass machinery with your child, pause to watch the drilling machines, street-cleaning machines, huge cranes, or cement-mixers. Your child may be unexpectedly entertained.

A walk along the Chicago River to look at all the boats and bridges is a lot of fun. Or, if you happen to find out about a nearby movie or television shoot, take a walk to the site. You may catch a glimpse of the costumes, dressing rooms, lighting and electrical equipment, or even a movie star.

A walk in the evening to a familiar spot can be a new experience for a child. Take your child for a walk in the dark to his or her school. The familiar surroundings may seem not-so-familiar in the shadows of the night.

Another walk your child may enjoy is a stroll down an alley. Of course, your child should always be in the company of an adult when exploring the discarded treasures, unusual plants, and tiny garages that may be found in some of the more charming alleys.

Speaking of discarded treasures, garage sales are often lots of fun for children. You might want to set a price limit, say $2.00, and let your child sort through the items deciding how to spend the money.

Another fun outing is a visit to your local firehouse. If you go on a nice day and you're lucky, the fire fighters may be relaxing outside and willing to demonstrate a slide down the pole to a reverent four-year-old, or to let your child touch their enormous fire boots and protective gear. You might want to call ahead, just to make sure they don't mind.

SOURCES: WHERE TO FIND UP-TO-DATE INFORMATION

Calendars of Events

Most museums and zoos have monthly calendars that you can request by mail.

Chicago Association of Commerce and Industry: Tourist information and gift shop at 200 N. LaSalle Street, Chicago, IL 60601, (312) 580-6900.

Chicago Magazine: A monthly publication with extensive listings of entertainment and restaurants and which includes a section on current programs for children—"Kidstuff."

Chicago Public Library Cultural Center: Call or write to be put on the mailing list for their monthly calendar of events. They're located at 78 E. Washington Street, Chicago, IL 60602, (312) 269-2900.

Chicago Sun-Times: Read the Friday "Weekender" and the Sunday "Culture" sections.

Chicago's Visitor Information Center: Located at 163 E. Pearson in the old pumping station.

Chicago Tribune: The Friday "Weekender" and the Sunday "Arts" sections are informative.

Illinois Calendar of Events: Available free from the Illinois Office of Tourism, (312) 917-4732.

Special Events Programs: Available from the Mayor's Office of Inquiry and Information, (312) 744-6671.

Telephone Information

Chicago Convention and Visitors Bureau: For information on current events, call (312) 567-8500.

Chicago Office of Fine Arts: For information about free programs and exhibits call (312) F-I-N-E-A-R-T.

Chicago Tourism Council: Located at 806 N. Michigan Ave., Chicago, IL 60611. Call (312) 280-5740.

Chicago Weather Report: Call (312) 976-1212.

CTA Transportation Information: To find out how to get somewhere on public transportation call (312) 664-7200. Be prepared to tell them where you are and where you want to go.

Sports Information: Call (312) 976-1313.

Theater Information: Call (312) 977-1755.

Time: Call (312) 976-1616.

Chicago Maps

B. Dalton Booksellers—at all locations.

Chicago's Visitor Information Center, located at Chicago Avenue and Michigan Avenue.

Kroch's & Brentano's, 29 S. Wabash Avenue, Chicago, IL 60603, (312) 332-7500; 516 N. Michigan Avenue, Chicago, IL 60611, (312) 321-0989; 835 N. Michigan Avenue (Water Tower Place), Chicago, IL 60611, (312) 943-2452.

Marshall Field's (bookstore—at this location only), 111 N. State Street, Chicago, IL 60602, (312) 781-4284.

Rand McNally Map Store, 23 E. Madison Street, Chicago, IL 60602, (312) 332-4628.
Waldenbooks—at all locations.

2
CHICAGO'S GREAT MUSEUMS

ADLER PLANETARIUM

1300 South Lake Shore Drive
Chicago, Illinois 60605
(312) 322-0304

Looking for a cool and entertaining spot on one of Chicago's baking-hot summer days? Pop out to the half-mile peninsula, where the planetarium sits, to relax in the cool breezes coming off the lake and attend the always-captivating Sky Show. Older children, in particular, love the planetarium because it ignites their imaginations by bringing to life the wonders of the universe. However, it is difficult to spend a whole day here, so many visitors combine the planetarium with stops at the Shedd Aquarium and/or the Field Museum. Or, if you don't want to over-extend yourself jumping from museum to aquarium to planetarium, combine an afternoon visit to the planetarium with a picnic or a concert at Grant Park.

The planetarium was founded in 1930 by Max Adler, whose dream was to bring the stars and planets down to earth. To this end, he constructed an "artificial sky" in Chicago. Although it was built more than 50 years ago, the planetarium remains extremely popular, and it gives children and adults a clearer perception of their place in the universe.

Besides the well-known Sky Show, the planetarium has three floors of exhibits covering astronomy, space exploration, and scientific instruments. Unknown to most Chicagoans, the planetarium also offers extensive course programs for both children and adults. Classes for children ages three to twelve include Star Dippers, Moon Maze, Ring Around a Planet, and Worlds for Sale. Another little-known planetarium special is the children's Sky Show shown every Saturday at 10:00 A.M. for children under six years of age.

Sky Show

The adult-oriented (and older child-oriented) Sky Show is the main event at the planetarium; frequently visitors ignore the exhibits and come just to see the show. Four new Sky Shows are written and produced every year, and depending on your interests and the show's topic, the experience is either fascinating and educational or just relaxing and a lot of fun. Leaning back in the specially-made, cozy chairs in the Sky Theater and looking up at the artificial, star-studded night sky on the planetarium ceiling are huge thrills for children.

The show is divided into two phases. Each one-hour Sky Show begins in the multimedia Universe Theater, where a 15-minute introduction sets the stage for the main presentation upstairs in the Sky Theater. Between 15 and 20 slide projectors are used at once during the introduction, and kids get to marvel at the world's largest map of the universe displayed on the ceiling of the Universe Theater. When the introduction is over, the audience leaves the Universe Theater and walks upstairs to the Sky Theater (elevators are available).

The 45-minute presentation in the Sky Theater uses the amazing Zeiss Mark VI planetarium projector, 100 to 150 special-effects projectors, and an argon-krypton laser, which all contribute to bringing the universe dramatically to life. The Zeiss projector is capable of putting the audience at any location on the earth at any time in history—past, present, or future. Kids are captivated as they are carried to the surface of the moons and planets of our solar system, to the center of the Milky Way galaxy, to the edge of a black hole in space, or to the distant realm of the quasars.

Sky Shows are shown Monday–Thursday at 2:00 P.M.; Friday at 2:00 and 8:00 P.M.; Saturday, Sunday, holidays, and all summer days at 11:00 A.M. and 1:00, 2:00, 3:00, and 4:00 P.M. Admission is $3.00 for adults; $1.50 for children ages 6–17; and free for seniors 65 and over with ID. Children under six are not admitted to regular Sky Shows (see Children's Sky Shows).

Main-Level Exhibits

The main level of the planetarium, which is really the basement, contains the exhibits most interesting for children. The Race to the Moon section chronicles America's manned space program. The center display island takes children back to a 1960s living room with a TV broadcasting highlights of the space program's beginning years. Nearby is a 1,015-pound meteorite, which children love to touch, from the Arizona Meteorite Crater.

The most exciting exhibit for kids is the area of futuristic Solar System Scales. The scales are little booths that let people know how much they would weigh if they were standing on the surface of the sun, moon, Mars, and Jupiter. Another interesting exhibit for children is Satellites: Watchers Over the Earth. This features an interactive game to identify Landsat photographs, a diorama of a space shuttle launching a satellite, and a hypnotizing pinball game that is a hands-on demonstration of a geo-synchronous orbit.

Mid-Level and Upper-Level Exhibits

The exhibits on the first and second floors are, for the most part, too sophisticated for children. The first floor has exhibits pertaining to historic telescopes, and the upper level concentrates on the various types of navigational instruments.

One neat sight children enjoy is on the upper level and does not pertain to the exhibits. On the west end are twelve bronze-framed glass doors. The heavy glass has been cut to form prisms around the perimeter of each pane. When the sun shines through the glass, the rainbow effect is spectacular.

Planetarium Store

Located on the main level, next to the entrance, the planetarium store is an excellent shop to fulfill the wishes of any child who has taken a particular interest in the stars. The store contains a wide variety of educational books; fantastic wall posters and pictures of awesome views of the universe; telescopes and other astronomical instruments and models; and an assortment of toys and games. Postcards of the lunar landscape and other out-of-this-world locations are available and fun to send to family and friends with a typical, "Having a wonderful time; wish you were here."

Children's Classes and Sky Shows

In classes at the planetarium, the kids are challenged by questions like: "Is Venus the morning star or evening star?" "Where's Saturn?" "If you lived in China a thousand years ago, would you see a Big Dipper or the Emperor's Chariot?" In a class called the Milky Way Mailbox, kids send postcards to alien kids asking them what they did on their summer vacations. All children's classes are held on Saturday mornings. One subject is offered per day and classes (about $10 each) are grouped together according to age. For more information on the upcoming schedule of classes write: Education Department, The Adler Planetarium, 1300 S. Lake Shore Drive, Chicago, Illinois 60605; or call (312) 322-0304.

The Children's Sky Shows are a special treat for little ones. One recent Star Stories show featured scary lions, beautiful princesses, and a friendly bear. Kids were also visited by Meteor Mouse, and they played connect-the-dots with constellations. Sky Shows are especially exciting for city kids, who rarely are able to see many stars at one time (even though the stars are artificial). All shows cost $1.50 per person, and parents are encouraged to attend with children. Show time is 10:00 A.M. on Saturdays.

Facilities and Access

Hours: The Adler Planetarium is open daily from 9:30 A.M. to 4:30 P.M. and on Fridays till 9:00 P.M. It is closed on Thanksgiving Day and Christmas Day.

Admission: Free, but the Sky Shows charge $3.00 for adults and $1.50 for children. The staff recommends that children under six do not attend regular Sky Shows.

Transportation and Parking: By car the Adler Planetarium is reached by Lake Shore Drive. By CTA, take the 146 bus, which can be caught along State Street. Metered parking is available on McFetridge Drive, and there is free parking next to the Field Museum (about a half-mile walk).

Restaurants and Food: Gulliver's Snack Bar on the main level has decent soups, sandwiches, and a salad bar that are all fairly priced, and there is a pleasant, circular dining area that gives people the feeling that they are eating in a crater of the moon.

Restrooms: Child-accessible bathrooms are available on each floor of the planetarium. There are changing tables in the main-level washrooms.

Information: Call (312) 322-0304 for personal responses or (312) 322-0300 for the general-information recording.

ART INSTITUTE OF CHICAGO

South Michigan Avenue at East Adams Street
Chicago, Illinois 60603
(312) 443-3600

Six-thousand-pound, bronze-casted lions guard the entrance to one of the world's finest collections of impressionist painting and classical sculpture. Although your children may want to spend the day climbing on these august, imposing animals, once you get your kids in the door, they will encounter an exciting and accommodating world. The Junior Museum, built and maintained solely for children's enjoyment, offers an assortment of hands-on, art-related activities designed especially for kids. In addition, the patient staff's primary responsibility is to respond to children's constant, wide-ranging inquiries.

Junior Museum

Where are viewing windows three feet above the floor? Where are tree trunks running through the middle of benches in the lobby? Where is there finger paint on sale at the gift store? And where are artists' life histories presented in simple, conversational English? The answer to all of these questions is the Junior Museum, which opens each day at the same time as the Art Institute, but closes 15 minutes earlier.

The Junior Museum, located on the basement floor directly below the main entrance and lobby area, presents art in a way that appeals to children's curiosity and sense of adventure. Next to replicas of impressionist paintings are simple fact sheets about the artists' lives. Alongside a painting by Monet, there is a picture of the artist and a caption explaining, "When was he born?" and "Where was he from?" Beside Seurat's *Afternoon on the Island of*

Le Grande Jatte is a blown-up reproduction of a detail in the painting, and through this magnified reproduction, a child can see the impressionists' "dot making," or pointillism technique.

Exhibits change from season to season, but the displays in the Junior Museum are always geared to the needs and interests of children. Artists often work in the entry hall of the Junior Museum, answering questions as they ply their crafts.

One popular child-targeted display is the Thorne Exhibit of Miniature Furniture now appearing in the Hammerman Gallery (one of two galleries in the Junior Museum). Therein, miniature rooms decorated with miniature furniture are displayed behind viewing windows only three feet off the floor. Beside the windows, children can peek into special holes in the wall that reveal people who might have inhabited these traditional rooms. As part of the Elizabethan-period exhibit, kids place their heads atop a cardboard cutout of traditional Elizabethan clothes. Then, by looking into a mirror, the children see themselves dressed in the traditional garb.

Within the Junior Museum there are numerous other interesting destinations. Note: the Junior Museum requires that children under eight be accompanied by an adult.

Little Library

Hours: Monday–Friday, 10:30 A.M.–2:00 P.M.; Saturday, 10:30 A.M.–4:30 P.M.; and Sunday, 12:00 noon–4:30 P.M. Storytelling, September–June 30, Sundays at 2:00 P.M. Staffed during the weekdays with a warm, knowledgeable corps of volunteers, the Little Library is an excellent source of art-history reference materials for adults and children. The librarians make an effort to display material directly relating to the special exhibitions in the adult galleries. Typical of the kind of dedication you'll find at the Little Library is Mary Doan, who has volunteered as a librarian for the past 18 years. Other docents have 20 or more years of service to the museum, and many specialize in children's affairs.

Junior Museum Gift Store

Located in the Junior Museum's main lobby, the Gift Store specializes in arts and crafts materials, such as fingerpaints; rainbow crayons and markers; safety scissors; spindle tops; the Junior Museum's own series of postcards; and T-shirts with the signatures of famous artists painted on the front.

In addition to the Junior Museum, the Art Institute also sponsors an activity-based Family Program (run by the Junior Museum staff) that acquaints children with the works displayed throughout the other galleries and the outside world.

Junior Museum Architectural Hikes

The Junior Museum offers five walking architectural and sculptural tours of the downtown area. Each of these self-guided walks is described in brief brochures written and photographed for children. Parents can obtain these brochures in the Little Library at no cost. All of the hikes originate from the lions in front of the institute.

Junior Museum Gallery Games

The Junior Museum sponsors a series of free Gallery Games (I Spy, Bits and Pieces, and Scrutinize) for children. I Spy, a game of pictured and written clues, takes children on a discovery mission to works of art all over the museum. Scrutinize involves the comparison of postcard pictures to the original works hanging in the main galleries. If you want to try one of these games, you can obtain materials and information in the Little Library.

Gallery Walks

Focusing on one specific artistic theme, Gallery Walks

(scheduled on Saturdays and Sundays) are special, guided tours of the Art Institute and its surroundings, offered to Art Institute members and their families. Certified docents from the Junior Museum lead tours based on a wide range of artistic topics, such as Native Arts, Rodin Revalued, and Gardens and Parks.

Family Workshops

Occurring on weekend mornings, these workshops consist of a brief discussion of a central theme followed by hands-on activities. Past themes have included: mini-model building, toothpick construction, sketching in the galleries, bookmaking, and origami.

Artist Demonstrations

Many children and families enjoy the artist demonstrations in the Art Institute. While working on projects in a public space, the artists welcome questions about their particular crafts and the process of creating art. Recent demonstrations have included watercolors, kite making, stenciling, and bronze casting.

For information regarding any or all of the above programs, please call the Department of Museum Education at (312) 443-3680. Ask for a current Family Programs brochure.

Facilities and Access

Hours: The museum is open every day of the year except Christmas Day. Monday, Wednesday, Thursday, and Friday, from 10:30 A.M. to 4:30 P.M.; Tuesday, from 10:30 A.M. to 8:00 P.M.; Saturday, from 10:00 A.M. to 5:00 P.M.; Sunday and holidays from 12:00 noon to 5:00 P.M. Guards begin closing galleries 15 minutes before the end of the day. The Junior Museum closes 15 minutes before the Art Institute closes.

Admission: Children under six, students with ID cards, senior citizens, and Art Institute members are admitted free. Tuesdays are free to all. On other days visitors may pay what they wish, but they must pay something. Recommended by the Art Institute: $5.00 for adults; $2.50 for children.

Transportation and Parking: The Art Institute is easily accessible by public and private transportation. The Adams Street el station stands one block west of the main entrance to the institute. In addition, any bus that travels on Michigan Avenue will stop in front of the lions. If you come by way of the subway, get off at Adams and State Streets: you will be two blocks west of the Art Institute entrance.

If traveling by car, the Art Institute staff recommends parking in the Monroe Street Garage ($5.50 for up to 12 hours). To arrive at this garage, simply turn off Lake Shore Drive onto Monroe Street, and the garage is on that first block. The Art Institute is one block south of the garage.

Restaurants and Food: The most popular (and economical) meal to have during the summer is a packed lunch at the picnic tables in Grant Park, near Columbus Avenue. For non-picnickers, however, there are three separate eating areas in the Art Institute: La Promenade, located between the School of the Art Institute and the Art Institute itself, is a small, cafe-style restaurant; The Garden Restaurant, open in the spring and summer months, is an outdoor affair where children can walk around and look at the fountain statuary while waiting for lunch; and the cafeteria is a smorgasbord-type.

Restrooms: Child-sized boys' and girls' facilities are just off the lobby area in the Junior Museum. Sinks stand about two and a half feet above the tile floor and the toilets are close to the ground. Regular-sized restrooms are located on the first floor of the adult museum.

Information: For daily information on a recorded message, call (312) 443-3500.

CHICAGO HISTORICAL SOCIETY

Clark Street at North Avenue
Chicago, Illinois 60614
(312) 642-4600

The Chicago Historical Society, located at the south end of Lincoln Park, is a privately endowed, independent institution devoted to collecting, preserving, and interpreting the history of Chicago, the state of Illinois, and selected parts of America's past.

The Chicago Historical Society is Chicago's attic, as the Smithsonian Institute is America's attic. And what a treasure trove it is for Chicago! The niches in the front lobby give testimony to the society's collection of more than 20 million objects. Most of the items on view in these niches were used or made in Chicago or depict noteworthy individuals from the city's past. Others epitomize Chicago's contributions to American history or reflect strengths in the society's permanent collection.

The Chicago Historical Society staff collects, preserves, and presents priceless artifacts and memorabilia from Chicago's honored past. This museum has outstanding children's exhibits and conveys a profound sense of Chicago's history for every child to absorb and treasure. A walk through the Historical Society will impart more than an armload of history books and is a lot more fun.

Newly remodeled and enlarged, the Historical Society now has ample space in which to display its wondrous collection. Ask for a monthly calendar to keep up with their changing exhibits. The permanent children's exhibits are unique and brilliantly executed.

Craft Demonstrations

Daily spinning, weaving, dyeing, quilting, candle dipping, flax processing, and printing demonstrations are held in the Illinois Pioneer Life Gallery.

Museum Store

The store sells a wide assortment of books on Chicago and Illinois; posters; postcards; Victorian wrapping paper; beautiful reproductions of the society's collection of 19th-century jewelry; T-shirts; corn husk dolls and toys; and exquisite beaded boxes. American Express, Mastercard, and Visa are accepted for purchases over $10.

Hands-On History Gallery

This large room is comprised of several areas, such as an old trading post with actual animal pelts that children can touch and old ledger books for children to explore and learn how barter and trade operated in early Chicago.

Another area is devoted to Chicago's early mail-order catalogs—like Sears, Roebuck and Montgomery Ward. The exhibit includes tape measures and explicit instructions on how to measure oneself for clothing offered in the catalogs. In this area children can even mount and pedal an old big-wheel bicycle; it's stationary and completely safe.

Children will love the old-time radio sound-effects exhibit, where they can handle and manipulate the devices, such as a barrel of blocks and coconut shells for kids to make the sounds of Fibber McGee's closet, a horse galloping, and people marching. There are about a dozen items here for a child to explore, and adults cannot resist joining in the fun, too.

After producing radio background sounds, a child can sit in an easy chair, turn on an old-time radio, and listen to radio advertisements from actual tapes.

Discovery Corner

Here arranged on shelves at a child's height are curious and enchanting boxes, each with a clue as to its contents. Children can take the boxes to adjacent tables and explore the contents—for example, a pair of high-button shoes, a mail-order blouse, or a bustle to try on. The children can have a tactile experience with items from the past as well as look at pictures and read articles that are included in the boxes. Few children leave until the contents of each box have been thoroughly explored.

A large Early American braided rug invites little ones to sit down and play with baskets of wooden blocks, Lincoln Logs, and other toys that must have amused our great-grandparents years ago. This is a room of wonder and delight, lovingly designed to enchant, surprise, and educate. It draws capacity crowds on the weekends, so you may want to visit during the week—either early morning or late afternoon to avoid school tours.

Chicago History Galleries

The galleries on the second floor contain an overview of major themes in the city's history: commerce, industry, transportation, culture, world's fairs, and neighborhood life. There are posters, buttons, bicycles, a huge teapot that looks as if it came from *Alice in Wonderland,* and a wonderful melange of other memorabilia, but center stage is the actual *Pioneer,* Chicago's first steam-engine locomotive, which ran between Chicago and Galena. An entire wall was removed to get the 12-ton locomotive into the museum. Children squeal with delight as they climb into the cabin and explore all the moving parts used to operate the engine. It is virtually indestructible. For a treasured photo of your child, position yourself, or a friend, in front of the locomotive and snap the picture as the child comes around the corner and views this iron behemoth for the first time!

Chicago's first fire engine, an 11-minute tape of the history of the great Chicago fire, artifacts from the fire including a treasured doll that was miraculously saved, as well as a listening booth with selections of early Chicago radio and music are some of the other exhibits in these galleries.

Illinois Pioneer Life Gallery

The pioneer gallery on the first floor is worth more than all the history books on Illinois. Here is a collection of pioneer tools and artifacts with actual demonstrations in natural settings of weaving, spinning, quilting, candle making, yarn dyeing—using dyes made from natural items such as onion skins and walnut shells, printing, corn milling, shingle making, and rope making. The two large looms are the oldest originals in Chicago. Museum docents perform all of the above pioneer tasks so the children see the work in progress. In the discovery section children can identify some of the early implements. Souvenir candles can be purchased in the bookstore.

In addition to the many changing and permanent exhibits, the society presents talks, seminars, craft programs, and music programs in the auditorium. You will want to obtain a floor plan when you enter and subscribe for the monthly calendar of events. Some of the 1989 programs include rug-braiding and quilting workshops, a tour of White Sox Comiskey Park—including a game against the Yankees, a Flag Day talk and exhibit, and a Fourth of July Celebration.

Facilities and Access

Hours: Monday–Saturday, 9:30 A.M.–4:30 P.M.; Sunday, 12:00 noon–5:00 P.M..

Library/Archives and Manuscripts Collection Hours: Tuesday–Saturday, 9:30 A.M.–4:30 P.M. Other research collections are open by appointment only. The society is closed on Thanksgiving, Christmas, and New Year's days.

Admission: $1.50 for adults; $.50 for children (ages 6–17); $.50 for senior citizens. Members free. No admission charge on Mondays.

Transportation and Parking: Metered parking one block north off Clark Street, and along Clark Street and

North Avenue. CTA buses 11, 22, 36, 72, 151, and 156 stop nearby.

Restaurants and Food: A new full-service restaurant is opening in the spring of 1989, although a picnic is always nice on the south lawn, which is quiet and secluded.

Restrooms: There are ample restrooms on the first floor, along with a checkroom and public phones. All exhibits and facilities are wheelchair accessible. There are both stairs and elevators to all floors.

Information: Call (312) 642-4600.

FIELD MUSEUM OF NATURAL HISTORY

Roosevelt Road at Lake Shore Drive
Chicago, Illinois 60605-2497
(312) 922-9410

Featuring tremendous dinosaur skeletons, totem poles, mummies' tombs, and exotic gems, the Field Museum is one of the world's best natural history museums. But beware! The Field Museum has more than nine acres of exhibits to see, hear, feel, smell, and explore, so it is virtually impossible to ward off becoming exhausted and mentally numb by the end of a visit. The key to avoiding fatigue is to plan your visit and to rest periodically. Explore a specific exhibit or subject (like the earth, plants, animals, or foreign people) thoroughly and don't worry about missing the others—the Field Museum is not going anywhere. Resting a few minutes beside the soothing water fountain in the central Stanley Field Hall while admiring the stuffed elephants or visiting the McDonald's in the basement will revive most fatigued adults and children.

Sizes

The new Sizes exhibit on the first floor is a fun first stop. This hands-on exhibit, designed to give children an understanding of the relativity of size and shape, is filled with thought-provoking entertainment. In fact, rumor has it that the exhibit is oriented toward so much "fun" learning, that some Field Museum purists have been grumbling that "the place is turning into a Disneyland museum." Regardless of old curmudgeon-like grumbles, don't miss the oversized kitchen table, where children can sit and experience what it was like to be two years old. Jump into a pair of size 67 blue jeans and try on the shoulder pad of William "the Fridge" Perry. The optical illusion room makes kids magically shrink and grow, and a TV screen shows Monster Mash movies. Also available is a reading corner with such favorites as *Tom Thumb, David and Goliath,* and *Jack and the Beanstalk.* And there are couches!

Place for Wonder

Next to the Sizes exhibit is the Place for Wonder, a must for all children. Open from 11:00 A.M. to 4:30 P.M. on weekdays, and from 10:00 A.M. to 4:30 P.M. on weekends, this two-room gallery is a hands-, eyes-, ears-, and nose-on exhibit. Greeting children at the entrance is Charlie, the life-size, growling polar bear. Children can enjoy having their pictures taken while sitting on a bench underneath Charlie's jaw. Next to Charlie are a big lizard, an elephant jaw, a stuffed raccoon, a huge shark fin, and numerous other touchable items. Many people overlook the excellent discovery boxes in the first gallery's desk drawers. Don't make that mistake. Ask a volunteer to show you the fur box (real buffalo, antelope, and rabbit furs inside) and the mineral box (fluorite, bauxite, copper, etc.).

The next gallery in the Place for Wonder is amazing: it is a miniature Field Museum on a child's scale, covering the four subjects of the museum: anthropology, botany, zoology, and geology. Touch a wooly mammoth tooth or a meteorite; smell natural vanilla and nutmeg; listen to a seashell; play with toys from other countries; try on costumes from the Orient; and introduce your child's senses to the wonders of the natural sciences. The Place for Wonder is staffed by excellent volunteers who are long-time experts in directing and explaining to children some of the more exciting objects on display. Consult these volunteers whenever you have the chance.

After experiencing the Place for Wonder, children should be more ready to tackle the larger and more complicated exhibits in the museum.

Foreign and Native Peoples

Across from the Sizes exhibit on the first floor is the Native American display containing exhibits on the Indians of South America, Central America, Mexico, the Southwest, the Plains and West Coast, and the Northwest Coast. The museum emphasizes that not all Indians are as homogeneous as many Americans think. Each tribe and area has its own culture and traditions.

The Pawnee Earth Lodge, a life-size re-creation of the tribal dwelling, is the center of the exhibit. The sounds of Indian ceremony allow visitors' imaginations to take them back to the environment of 19th-century Pawnee. Check in at the information booth (in the center of Stanley Field Hall) upon arrival at the museum to find out the lodge's daily show times. The show—conducted by knowledgeable volunteers, anthropology graduate students, or staff—lets visitors sample the daily life of the Pawnee Indians. If you miss show time, or if the performance is booked, you can still look into the beautiful Earth Lodge and listen to an audio recording of the show, but it is not nearly as interesting or exciting.

Near the entrance of the Native American exhibit is a selection of vivid Indian costumes and a display of cultural masks intended to shock the viewer. Your children will probably like the dramatic and wild-looking masks with huge lips, distorted expressions, and long witch-like hair. Even a child can feel and see the masks' impact and know how difficult it must have been to make these fantastic art pieces.

To the right of the lodge is a beautiful forest of towering totem poles, each bearing the maker's history and expressing his religious and cultural sentiments. Throughout the Native American exhibit are video pods featuring dances and religion, but unless you are accompanying a budding archaeologist or unless you need to sit and give the baby a bottle, we don't recommend stopping for them: they are a bit dry for most children.

Behind the forest of totem poles is a comprehensive exhibit on the Native American Eskimos and the Northwest Coast Indians. The exceptionally crafted, low-to-the-ground displays reveal the secrets, traditions, and survival techniques of these people.

To the left of the Earth Lodge is a more traditional museum exhibit focusing on Pre-Columbian Indians. And if you or your children have any interest in furthering your studies of Native American culture, do not miss the new Webber Resource Center, also to the left of the lodge. The center contains resource material on Indian life, and the staff can direct you to books or magazines relating to your specific interests.

On occasion, the museum staffs roving exhibits of Indian games that offer the tactile experience of Indian children's play. Each game strengthened a skill required for the Indians' survival.

Animals

The mounted animals next to the Place for Wonder galleries are so lifelike you might think you see a nose twitch or an eye blink. Displayed in diorama form, the animal exhibits range from insects to elephants, from Africa to the Antarctic, from skeletons of a black right whale to a tiny short-tailed shrew. These dioramas offer a clear comparison of the evolutionary relationships among species.

The animal section on the main floor is divided into seven parts: Mammals of the World, Mammals of America, Mammals of Asia, Mammals of Africa, Skeletons and Reptiles, Bird Habitats, and Birds. The thousands of species provide the possibility of many games. One game that kids like is finding which species is the heaviest, which is the biggest, which is the fastest, which is the smallest, etc.

Plants

The Plants of the World Hall on the second floor is a spectacular collection of thousands of plant models handcrafted petal by petal, and some even seed by seed. See the largest leaf in the world or the microscopic diatoms and algae; view the fig leaves and see if the children could clothe themselves with them.

Flowers, fruits, leaves, stems, roots, and seeds are arranged according to plant families, and many exhibits include magnified and cutaway views of the interior plant parts. Full-scale dioramas depicting an Illinois forest of wildflowers in the spring, an alpine meadow, an Amazon water lily habitat, and algae in the North Atlantic coast are also included in the hall. In an adjacent area, hundreds of plants of economic importance are featured along with models and representatives of their products. While not ordinarily intriguing, the Plant Halls are often uncrowded and provide great roaming room for restless toddlers.

The Earth

In the geology halls, rocks and minerals from around the world dazzle with a rainbow of colors and even glow in the dark. Located on the second floor, the precious gems and jades galleries contain not only everyday materials, such as salt, copper, and coal, but also out-of-this-world meteorites and a relief map of the moon. The geology exhibits cover everything from re-creations of prehistoric seas, where life dawned, to the evolution of some of the largest creatures that ever lived. The halls have beautifully dark settings, perfect for geological and aesthetic treasures.

The Dinosaur Hall, the all-time child pleaser, is right next to the gallery of gems, and it is anchored at one end by the famous diorama of a Carboniferous Swamp Forest that is the home of the 72-foot anatosaurus, one of the great dinosaur skeletons at the museum. *Danny and the Dinosaur*—the story of a little boy's dream of the perfect pet—takes on new meaning when a child views the great size of these beasts.

Field Museum Gift Shop

This is one of the best museum shops in Chicago. It has beautiful woven baskets; bead necklaces; hand-carved sculptures; sweatshirts; unusual stuffed animals, such as giraffes, zebras, and dinosaurs; and a thorough collection of books and music cassettes. If you have any spare pocket change, make this a stop before you leave the museum.

Classes

Unknown to many Chicagoans, the Field Museum offers an excellent array of classes for youngsters with a particular interest in natural history. The classes, geared to children ages four to thirteen, are given year-round but are more numerous in the summer.

So if you want your child to hear some alligator stories or to investigate the Mayan mysteries, write the Field Museum of Natural History, Department of Education, Summer Fun 1989, Roosevelt Road at Lake Shore Drive, Chicago, Illinois 60605-2497.

One special event at the Field Museum is an overnight adventure for nine- to ten-year-olds. It's called, "A Night in the Field," and it is just that—a trip behind the scenes. Included is dinner, a movie, and camping out in "The Field" for the night.

Facilities and Access

Hours: The Field Museum is open from 9:00 A.M. to 5:00 P.M. daily. It is closed only on Thanksgiving, Christmas, and New Year's Day.

Admission: Adults, $3.00; ages 2–17, students, and seniors, $2.00; children under 2, free; families, $10.00; and on Thursdays, all admission is free.

Transportation and Parking: The museum, located next to Soldier Field, can be reached by Lake Shore Drive or by CTA bus 146 available on State Street. Large lots are conveniently located next to the entrances.

Strollers: Available for a $1.00 refundable deposit at the checkroom on the first floor.

Restaurants and Food: There is a large McDonald's located on the ground floor open from 9:00 A.M. to 4:30 P.M. daily. (It's a bit noisy, but it's OK.) A large snack area with vending machines is next to it and is open all day.

Restrooms: Child-accessible washrooms are available on every floor. There is a special changing and feeding room located on the ground floor next to the Ancient Egypt exhibit.

Information and Reservations: Call (312) 922-9410.

MUSEUM OF SCIENCE AND INDUSTRY

57th Street and South Lake Shore Drive
Chicago, Illinois 60637
(312) 684-1414

Children enter the Museum of Science and Industry wide-eyed and energetic. Immediately, they are hit with the impulse to dash off and push every button, crank each handle, pick up all the phones, and play with everything the museum has to offer. But wait. Grab a hold of your children's shorts before they scatter in all directions. You are dealing with the Museum of Science and Industry—the granddaddy of Chicago's museums, and the leading tourist attraction in the Midwest. Over four million people pass through the doors each year, and too many simply wander through the 90-plus exhibits without direction or focus. The Museum of Science and Industry—as all other mega-museums in Chicago—is more enjoyable if you plan ahead.

First, decide what day to go. Mondays, Tuesdays, and Wednesdays are less crowded than the other days of the week. The museum is open on all holidays and is not crowded on Thanksgiving Day morning or New Year's Day morning. Another factor to consider is the time of the year. The museum is much less crowded in September, October, January, February, and March than it is during the other months of the year. Also, consider the time of day: if you want to avoid the long lines for the coal mine or submarine exhibits, arrive early (the museum opens at 9:30 A.M.).

The next step is to choose what exhibits your children would most like to see. There are some not-to-be-missed stops, including the Whispering Gallery, the Coal Mine, and the Omnimax Theater, but remember, these are the most popular and crowded exhibits. The variety of things to do and see at the museum is staggering, and there is no need to follow the crowds to the most popular places.

Go to the Food for Life exhibit and see hatching chicks or walk through the giant heart. Head to the Grainger Hall of Basic Science to watch some science demonstrations. If you have a child under six, reserve a spot upon arrival at the Curiosity Place so he or she can spend an hour exploring light, motion, force, and sound. Or choose a subject you and your children know little about and visit those exhibits relating to it. If you are still pondering where to go, pick up the Self-Guided Tour brochure at the Information Booth and follow its excellent suggestions for the day.

Hot Spots

Certain exhibits in the Science and Industry are surefire hits for kids and adults. However, none is so fantastic that you should wait an hour in line for it. Schedule your trip to avoid the midday rush for popular exhibits. Early mornings and late afternoons offer the shortest lines. Here are the top six spots:

Coal Mine: The traditional Science and Industry favorite (located in the South Court on the entrance floor) is a realistic reproduction of a typical southern Illinois coal field. The tour is an exciting descent through the main shaft in a darkly lit freight elevator that feels as if it were traveling a mile beneath the earth. Everyone is herded onto a loud screeching electric train to go to the working "face," or coal vein, where children can enjoy the thrill of being in a strange, new, and creepy environment (perhaps more than listening to the guides' description of the techniques of coal mining). The tour concludes with a literal "bang" as a guide demonstrates the dangerous explosive power of methane gas in a protective container. Admission is $1.50 for adults, $1.25 for children 12 and under.

U-505 Submarine: This submarine (located on the ground floor, east) is an actual German submarine captured during World War II. The conducted tour is followed by a film depicting the 1944 capture. Seeing the cramped quarters of the submarine fascinates kids, who find it difficult to believe that people lived in such a cooped-up environment. Like the mine, the actual atmosphere of the submarine is more enthralling than the explanations of the technology used for living under water. Admission is $2.50 for adults, $2.00 for children 12 and under and seniors.

Omnimax Theater: With a five-story domed screen and 72-speaker sound system, you not only watch the film, you feel like you are in it. The films, which change about every three months, are visual spectaculars. Past films have included a Space Shuttle flight, a look at the Great Barrier Reef of Australia, and a rafting trip down the Grand Canyon. Located on the entrance floor in the East Wing, the theater schedules shows every 50 minutes. Cost: $4.75 for adults and $3.25 for children and seniors; be sure to request tickets in advance by calling (312) 684-1414.

Model Railroad: Another traditional favorite at the museum is the model railroad. The exhibit seems to hypnotize youngsters, who love to stare from the balcony above and wander the perimeter on the entrance floor. The

3,000-square-foot maze of trains, cities, farms, and mountains is every child's dream Christmas present: it's a train set the size of a house. The trains are controlled from a switch box at the end, which is always manned by a guide-lecturer who answers questions concerning its operation. This is an excellent place for fatigued parents to catch their breath on a nearby bench while easily keeping their eyes on their train-bewitched children.

Whispering Gallery: Probably the most memorable of exhibits for young children, the Whispering Gallery amazes and titillates. The shape of this hall allows sound waves to travel from one end to the other, so children standing at one end can clearly hear a whisper from the other end. Kids experience a deep sense of shock and wonder when they hear their friend or sibling or parent whisper hello to them from over 30 feet away. Exhibits along the wall explain the nature of sound and electromagnetic waves through the use of models. A bell jar demonstration proves—audibly—that sound cannot travel in a vacuum.

Yesterday's Main Street: Flickering gas lamps and old store fronts depict a walk down a Chicago street in 1910. A cinema shows silent-film classics for $.25, and Finnegan's Ice Cream Parlor provides an unexpected resting spot and highlight, with sodas and sundaes that are reasonably priced.

13 More Sure Hits

Circus (entrance floor, East Pavilion): Thousands of hand-carved miniatures set in animated scenes recall the history and excitement of the tent circus in America, the circus parade, tent layouts, "under the big top," and side shows presented with narration. Kids love to activate special colored lights, which transform their own reflections into clowns. An eight-minute film of circus performers shown on a 30-foot vertical screen presents a view of circus life and acts. The film is quite popular.

Classic Cars (ground floor, East Pavilion): Are your children car buffs? Then bring them to this exhibit to see the 16-cylinder Cadillac, Rolls Royce, Brewster, Duesenberg, and other classic models.

Curiosity Place (mezzanine level, yellow stairwell): This is a must-stop for children under six. The exhibit is divided into four sections: (1) the basics of light are presented through a two-way mirror, a walk-in kaleidoscope, and a shadow screen; (2) an introduction to the world of sound is provided by musical instruments, a xylocoaster, speaking tubes, and a tone tower; (3) the concepts of force, machine, and motion are represented by the exhibit's child-powered crane, inclined planes and pulleys, double swings, and sand pendulum; and (4) fun water play is featured. Throughout the exhibit, preschoolers are physically involved in the process of simple problem solving to help them learn fundamental scientific principles. Reservations are required for groups of 20 or more. Call (312) 684-1414, ext. 326. Curiosity Place is open 10:00 A.M.–5:00 P.M. in the summer and 10:00 A.M.–3:30 P.M. the rest of the year. It's closed on Tuesdays.

Energy Lab (ground floor, yellow stairs): Energy in its various forms is the theme of this exhibit, and the individual displays highlight a specific conversion of one form of energy to another. Check out the solar energy section first. There is a visitor-activated model of the New Mexico solar power tower experiment and an interesting description of the museum's solar heating and cooling system. Be sure to see the Rube Goldberg–type device, located in the center of the exhibit, which imaginatively combines simple mechanical principles to convert a ball's gravitational potential energy to mechanical energy. This exhibit is best for older children interested in science.

Fairy Castle (ground floor near yellow stairs): If you have a little one, stop by this wonderful dollhouse, commissioned by silent-film star Colleen Moore. The castle contains jeweled depictions of international fairy tales, such as *King Arthur, Robin Hood, Robinson Crusoe, Aesop's Fables, Mother Goose,* and *Cinderella.*

Food for Life (entrance floor near blue stairs): An outdoor farmers market is the setting for this educational and entertaining exhibit. Plastic models of food and signs explain the five basic food groups and the necessity of each for good health. The big thrill of the exhibit is watching an incubator, where brave little chickens emerge from their shells wet and exhausted and turn into fluffy little peepers.

Other activities children enjoy are peddling a bicycle that measures the amount of calories they burn while exercising and choosing a menu to see if their choices are nutritionally balanced.

Actual experiments on plant nutrition are carried out in a live greenhouse area, and sample food tests are conducted at the Consumer Research Center, whose staff frequently seeks input from children and may want to interview your little ones. A special computer is located in the front of the exhibit to provide personalized nutritional information and answers to questions you may have.

A computer also allows your children to select their make-believe meals from revolving tiers of plastic goodies and then analyze whether the selection is healthful. It's fun to pick foods for both healthy meals and junk-food meals.

Foucault Pendulum (blue stairwell): This amazing three-story-high hanging pendulum duplicates Jean Foucault's 1851 proof that the earth really does rotate. The museum uses a clock face beneath the pendulum to show the motion of the earth throughout the day.

Grainer Hall of Basic Sciences (balcony level, west of blue or green stairs): One of the museum's largest and most popular exhibits, the Grainer Hall is divided into four main sections: Beginnings, Life, Fundamentals, and Inquiry. Computer games and simulations accompany many of the exhibits and are ideally suited for older children with a desire to tinker with computers. There's a sweeping range of activities: cell structure is explained through a creative matching game; a transparent model of a fish demonstrates internal structures; a film from the popular Cosmos program introduces evolution; gravitational force is described with a coin and feather drop experiment; and a great film, *The Powers of 10,* explains objects of large and small dimensions.

Here's Your Heart (balcony level, north): The famous 16-foot, walk-through heart is located here. The thump alone is worth a visit. Kids are intrigued by a push button that activates a lighted diagram showing the process of both a stroke and a heart attack. A mock-up of an operating room shows a patient hooked up to a heart-lung machine. Near the giant heart, in an adjoining room, is the Heart Kitchen, where push-button quizzes answer questions and indicate wise eating habits.

The Money Center (entrance floor): Here kids can play a pinball game showing the multiplier effect; respond to quiz questions, which, depending on the answers, cause an increase or a decrease in their exhibit account balances; and participate in various other games that give them a general understanding of economics. Other exhibits here include a case of artifacts that compares the buying power of today with the "good old days," a diagram of what happens when you write a check, and an explanation of how a buyer's choice of goods "employs" the supplier of these goods.

Space Exploration (entrance floor, East Wing): The actual module of the *Apollo 8* mission to the moon—the first to circle the moon—is on display: exciting stuff for any 10-year-old. Nearby models of the *Gemini* and *Mercury* spacecrafts offer an interesting comparison. Photographs, drawings, printed material, and models explaining the technology and aims of the space shuttle program fascinate the would-be astronaut in each viewer.

A Trip Through Time (entrance floor, East Court): The concept of time in the past, present, and in other societies is discussed through words and pictures. Diagrams, charts, and text explain the concepts of internal clocks, jet lag, and natural rhythms. Kids like the test that allows them to guess the length of one minute and the game that asks them to perform various tasks under pressure. This is another good spot for older children.

Wheels of Change (entrance floor): Using 42 rotating animated dioramas, American Crossroads shows how the automobile changed the life and landscape of rural America between 1900 and 1940. Kids love the 3-D effects of the movie, *The Wizard of Change,* which traces the evolution of man's tools, and car enthusiasts are enthralled with a look at futuristic designs for automobiles.

And Much, Much More

There are numerous other worthwhile exhibits to experience at the Museum of Science and Industry. All museum visitors should pick up a complete listing and description of exhibits at the Information Booth located right near the entrance. Further, the museum always has a few extraordinary, temporary exhibits on display. Call (312) 684-1414 to learn about the museum's current features.

The museum presents an excellent array of classes, weekend workshops, and mini-camps for children. Exciting, week-long Robotics and Space Day Camps are offered in the summer for kids ages 10–14. For children over eight, there are great one-hour parent-child workshops covering topics ranging from periscopes to heart beats. And there are year-round classes and hands-on activities for preschoolers and younger kids. Topics include sound, motion, electricity, and numerous others. For more information call (312) 684-1414, ext. 421. Or write the Education Department, Museum of Science and Industry, 57th Street and South Lake Shore Drive, Chicago, Illinois 60637-2093. The museum also sponsors teacher workshops to aid in the teaching of sciences.

Museum Store

Located in the rotunda and the Henry Crown Space Center, this large store offers slides, film, jewelry, models, toys, posters, books, T-shirts, and a lot more. Prices are reasonable and a great variety of items are available.

Facilities and Access

Hours: Museum is open every day. Hours vary by season: Memorial Day through Labor Day: 9:30 A.M.–5:30 P.M. For the rest of the year: 9:30 A.M.–4:00 P.M. on Monday–Friday, 9:30 A.M.–5:30 P.M. on Saturday, Sunday, and holidays.

Admission: Free.

Transportation and Parking: By CTA bus, el, and train. Free parking.

Restaurants and Food: The museum offers seven food services, which range from snacks to light lunches to full lunches to ice cream.

Restrooms: All located on the ground floor, opposite the Century Room.

Information: Call (312) 684-1414, ext. 2220.

SHEDD AQUARIUM

1200 South Lake Shore Drive
Chicago, Illinois 60605
(312) 939-2426

Stepping into the John G. Shedd Aquarium is like entering a Jacques Cousteau world of underwater surprise and enchantment. The 90,000-gallon Coral Reef Tank located in the front and center of the aquarium is the first sight to greet visitors upon their arrival. This exhibit has an exciting mixture of giant turtles side by side with swordfish, colorful blowfish, and a multitude of other sea animals. The main event at the aquarium is the Coral Reef Show

(11:00 A.M. and 2:00 P.M. daily; and a 3:00 P.M. show weekends) featuring a child-dazzling, deep-sea diver wired with a microphone who swims amongst the sea creatures. Watching him swim, feed the fish, and tell jokes is breathtaking for little ones. Even hearing him breathe is a special thrill (a spooky-sounding "blub, blub, blub").

Take Your Time

While the Coral Reef Exhibit is a can't-miss stop, the aquarium has numerous other attention-grabbing exhibits and galleries; in fact, there is so much to see and gawk over (more than 800 kinds of aquatic animals), one should be wary of becoming glossy-eyed from the overwhelming number of sights. Take your time and enjoy some of the extraordinary fish and animals rather than getting caught up in children's temptations to rush through trying to see everything at once.

The aquarium is centered around the Coral Reef Tank and all six of the main galleries jut off from this exhibit (all the exhibits are located on the first floor). Each gallery has its own special attractions. Gallery One contains bright green moray eels, lionfish, guitarfish, and small sharks; Gallery Two has wild-looking flashlight fish; a funny-looking leopard shark and beautiful starfish are in Gallery Three (along with an education exhibit); Gallery Four features the sea otters; catfish and turtles are displayed in Gallery Five; and Gallery Six contains some tremendous paca fish.

Activities Center

In addition to the Coral Reef Tank and the galleries, the aquarium has an Activities Center, a hands-on exhibit that is great for children of all ages. Here kids can touch and see shark's teeth, spike-shell blowfish (dead, of course), exotic crab shells, and other fun sea objects. Live specimens, placed under a microscope and blown up for a TV screen, capture the attention of most electronic-age kids and teach them a little biology. One problem with this exhibit is that the activities need to be supervised by aquarium volunteer staff. When there is a shortage of volunteers, the exhibit is closed. Try the mornings, when more volunteers are working, if you want to catch the Activities Center when it's open.

Tours

Behind the Scenes tours are another great way to enjoy the aquarium. If your children are old enough to be good listeners and have inquisitive minds, the tours can be worthwhile. The volunteer guides explain to children and adults alike the various traits and habits of the sea creatures. The tours also provide a nice structured way for seeing the aquarium galleries for those who don't like to wander. There is a $2 charge per person, and a minimum of 10 people are required. To schedule a Behind the Scenes tour, call the Volunteer Department at (312) 939-2426, ext. 378.

Depending on the time of year, there may be a special exhibit specifically designed for children. Occasionally, classes, shows (with puppets), and films are shown in conjunction with an exhibit. Since the aquarium schedule of events is constantly changing, the best way to discover if there are any new exhibits is to call Christine Westerberg (Special Exhibits Coordinator) at (312) 939-2426, ext. 337.

Everyday, however, the aquarium screens *Introduction to the Aquarium,* a short feature, every half hour on the hour and half hour in the main auditorium near the entrance doors. This film is an overview of the aquarium and can provide a rest opportunity for both tired parents and tired children to sit down and relax.

Gift Shop

The Sea Shop, a handy souvenir shop located next to the entrance, has great stuffed animals, aquatic books, earrings, and necklaces along with a collection of souvenirs that are not piggy-bank busters.

Classes

Unknown to most aquarium visitors, the basement contains an Aquarium Science Center, which houses spacious, well-equipped, educational facilities consisting of laboratories and classrooms. The aquarium conducts classes here for children ages 3–17 that round the gamut of subjects from sharks and whales to coral reef ecology to farming the sea. For more information on these superb classes, write to Tom Lincoln, Public Program Coordinator, John G. Shedd Aquarium, 1200 South Lake Shore Drive, Chicago, Illinois 60605, or call ext. 359.

Facilities and Access

Hours: The aquarium is open from March through October from 9:00 A.M. to 5:00 P.M. and from November through February from 10:00 A.M. to 5:00 P.M. It is closed Christmas and New Year's Day.

Admission: $3.00 for adults; $2.00 for seniors and children ages 6–17; and free for children under 6. Thursdays are free for everyone.

Transportation and Parking: The aquarium is accessible by Lake Shore Drive and the 146 bus. Park in the Field Museum lot (free) or along Solidarity Drive (metered), but be prepared—parking is far away from the aquarium and the walk, for most of the year, is cold and windy. Dress warmly, bring the stroller, and be prepared for the hike. Access is only via an underpass from the parking lot just north of the Field Museum. If that lot is full, the hike from behind that museum to the aquarium is over one-quarter of a mile. The entrance marked "Handicapped Only" is also available for those with strollers, even though it is not marked as such. Enter there to avoid the difficult task of carrying the stroller up the enormous staircase in front of the aquarium.

Restaurants and Food: There is no restaurant at the aquarium, and food and drink are prohibited. The closest one is in the Field Museum basement, a good half-mile hike door to door, and you must go outdoors to get there. So feed the children before you go. However, some benches in the front entrance are often filled with people feeding babies.

Restrooms: Sufficient restroom facilities exist only on the ground floor, one long flight of stairs down from the exhibit, but on the same level as the stroller entrance. The women's restroom includes a nice tabletop infant-changing area. There are no paper towels, only air-blowing machines, so bring your own baby-cleaning necessities. There is a drinking fountain outside the bathrooms.

Information: Call (312) 939-2426. Recorded message at (312) 939-2438.

3

CULTURAL GEMS—CHICAGO'S SPECIALIZED MUSEUMS

EXPRESS-WAYS CHILDREN'S MUSEUM

435 East Illinois Street
Chicago, Illinois 60611
(312) 527-1000

When you mention a "visit to the museum" to your children, do they scrunch up their noses? Do they think of a museum as a place where adults "ooh and aah" over lifeless bones and pottery shards while they try to figure out how to pry open the glass display case? Do they dread yet another slide show that features penciled diagrams, close-ups of skulls, elevator music, and explanations that cause adults to nod their heads when they can't even fathom what the narrator is talking about? And do you resent dragging your children by the hand down musty hallways and corridors simply to have them turn to you and say, "But we just saw something like that in the last room!"?

If you answer yes to any of these questions, then visit the Express-Ways Children's Museum, which resembles a sandbox more than a museum. Opening its doors in 1982, Express-Ways is an interactive and participatory museum where children learn by doing rather than by passive viewing. The hands-on exhibits and workshops, which focus on serious artistic and historical topics, are designed to stimulate creativity while also encouraging sensory learning and problem solving. Hanging on the walls of the museum are signs that straightforwardly communicate the Express-Ways philosophy. The signs say, "While at the Museum: Touch, Explore, Play!, Browse, Ask?, Read!, Use bathrooms for diaper changing."

Key Features

Touchy Business: Touchy Business, recommended for children ages 2–7, is a room full of sensory adventure for children. After removing their shoes at the entrance, kids play in and around several main activity centers, exploring and manipulating the various shapes and textures they encounter. There are several wooden benches for parents and grandparents who can't quite keep up with the frenetic pace.

Tactile Tunnel: Wearing blindfolds supplied by the museum, children crawl through the tunnel identifying ordinary objects (i.e. plastic tubing, telephones) by using their sense of touch. They also feel what's inside the "mystery jars."

Three Bears' House: Kids step into a miniature house of the three bears, complete with dining table and rolling window that changes the view outside from morning to afternoon to nighttime.

Amazing Chicago: Amazing Chicago, recommended for children ages 7–13, makes famous Chicago architectural structures accessible to children. At the miniature Art Institute building, kids create portraits of themselves with the aid of a mirror and a framed felt board. Inside the post office replica, children play with the postage meter and mail slots as they sit at a desk behind grates. Everything in and around this room is designed to stimulate children's interest in the city: pictures of famous buildings and architectural styles line the walls; a *Chicago Tribune* newspaper dispenser stands in the middle of the room; a child-sized CTA bus waits outside the entrance; and even the stools in the room have trivia questions written across them (i.e. when did the first el train begin running in Chicago?).

Magic and Masquerades: Magic and Masquerades, recommended for ages 7–13, literally brings children eye-to-eye with African culture. Displayed just a few feet off the floor are photographs of African villages, statues, board games, and dress. While looking at pictures of Andinkra shawls and native jewelry, kids piece together their own necklaces and bracelets with recycled scrap materials. Another highlight in this room is the miniature African hut, complete with drums and straw matting on the floor.

Recycle Arts Center: The Recycle Arts Center is a room stocked full of materials for making masks, necklaces, sculptures, or just about any type of low-budget art object you can imagine. A favorite room for teachers or parents of large families, it contains boxes and baskets full of foam rubber, wooden two-by-fours, metal coils, plastic ribbon, yarn, styrofoam packing material, etc. The museum charges $1.50 for a lunch bag full of materials and $3.25 for a shopping bag full. Even if you don't care to purchase anything, let your child stick his or her hands into the baskets and feel the different textures. Watch the delight of your kids as they play with styrofoam packing material: the texture is so light and fluid, it feels like letting sand run through your fingers.

Children's Workshops: Express-Ways offers numerous one-day, one-week, and weekly workshops for kids of all ages. Typically, these workshop sessions last about two hours and they focus on a specific artistic or cultural theme. Past themes include: Art that Goes (kinetic art); City Stalker, an adventurous trip around the Loop; and Behind the Scenes at the Teddy Bear Factory, a tour of the place where Teddies are "born"—children receive a free "orphan" bear and a birth certificate. The cost of these workshops is normally $3–$5 per daily session. For information regarding upcoming activities, call and request Express-Ways' most recent program listing.

Outreach Programs: Express-Ways has an extensive outreach program for schools, churches, libraries, or any interested large groups. Partially funded by the federal government, these activity-oriented traveling exhibits include Thunderations—Sounding Off About Dinosaurs; Good Fibrations, an introduction to the world of fibers; and Color Forms, an exhibit for toddlers that examines shape, form, and texture. Express-Ways staff accompanies the exhibit and facilitates the activities. These outreach programs have varying costs and they require two- or three-day minimum rentals. Interested parties should call the outreach-program staff.

Teacher Training: Express-Ways also offers workshops especially for teachers. These workshops are designed to show how hands-on art activities relate to reading, writing, math, and science. Past workshops have covered the following themes: Puppets with a Purpose, Bubbles and Simple Sciences in the Classroom, and Renoir in the Classroom. Interested people should call the teacher-training staff.

Gift Shop: The front desk sells a variety of children's items: calendars that feature children's art, wooden blocks with marble chutes, and T-shirts printed with the Express-Ways logo (a rainbow-colored handprint).

Facilities and Access

Hours: Tuesday–Friday, 12:30 P.M.–4:30 P.M.; Saturday and Sunday, 10:30 A.M.–4:30 P.M.; Monday, closed to the public. Special exhibit for toddlers open Wednesday, Thursday, and Friday, 10:00 A.M.–12:30 P.M.

Admission: Admission is $3.00 for adults, $2.00 for children; Thursday evenings 5:00 P.M.–8:00 P.M. are free.

Transportation and Parking: The metered spots are usually taken in this downtown area, but there are several pricey lots nearby.

Restaurants and Food: While there is no place to buy food inside Express-Ways, there are several eating options within the glitzy new North Pier Terminal—the new East Side of Chicago's Loop.

Restrooms: There are adult-sized boys' and girls' restrooms; both provide changing tables.

Information: Call (312) 527-1000.

Strollers: Express-Ways does not allow strollers inside the building. If you do bring one, though, don't worry. The front desk will provide you with a lock and cable with which you can chain your stroller to the metal rail on the front steps.

Membership: Families can obtain memberships to Express-Ways Museum. Benefits of membership include: free entrance year-round, discounts on purchases at the front desk, and the opportunity to hold birthday parties at the museum.

KOHL CHILDREN'S MUSEUM

165 Green Bay Road
Wilmette, Illinois 60091
(708) 251-7781
(Note: use the 312 area code before 11/11/89)

Children instinctively know that the best way to learn is by doing, feeling, and experiencing. The Kohl Children's Museum provides this ideal environment for exploring and learning. Every exhibit is an interactive environment that stimulates excitement and curiosity. Want your children to understand the life of an animal? Go to the Kids and Pets section where children wear animal costumes and use their noses to direct themselves through a special animal tunnel.

At the Kohl Museum there are no ropes, no glass cases,

and no signs telling kids, "Do Not Touch." The sneaky thing about the museum is that kids think it's just an indoor playground. Through play, however, children assimilate worthwhile concepts. It's exciting to watch them absorb the atmosphere of fun learning and discover some wonderful new ways to look at things.

The museum is targeted to ages two through ten, but older kids will find a few appealing activities, too. Call (708) 251-7781 in advance to learn about the daily special activities, such as puppet shows, sing-alongs, story times, or special guests. The wonderful array of exhibits also makes the museum worth a visit.

Who Am I?

Get the creative juices flowing right away by going to the Who Am I section, where children enter the world of make-believe. Here kids can paint their faces and imagine being anyone they please through different role-playing and dress-up games and performing in front of the video camera. Some children remain in face paint throughout the day. The kids seem to enjoy playing and seeing their friends in these new faces.

Kids and Pets

This area effectively depicts the balances, pleasure, and necessary obligations involved in owning a pet. The exhibit is instructive in that it displays the best way to care for pets and how to choose one, yet it is experiential in that it shows kids what it feels like to be an animal. This is a good way to explore the issues of pet ownership, for it gives children a taste of the caretaking a pet requires and the rewards of an animal's dedication to a child.

Bubblemania

This is the place for pure hands-on enjoyment. Children can try out the new bubble box and blow big, bouncing bubbles. The exhibit entertains the senses. You can see the colors, feel the bubbles, and listen to all the laughing and delighted giggles as children create enormous bubbles in wild shapes and take turns standing inside a gigantic soap bubble.

The Learning Locomotive

Preschoolers glory in this area, designed specifically for their levels of development. The toys available include blocks and marbles, Brio construction, and train sets, as well as preschool-appropriate computers and educational software.

Walk Through Morocco

Children can become part of Moroccan tradition and lore for a day, trying their hands at creating brass rubbings or traditional babooshes (shoes) and hamsas (good-luck symbols). They can also experience a different kind of daily family life in a traditional Moroccan living room and actually feel a culture that rests thousands of miles away.

Walk Through Jerusalem

Here color and richness of the old walled city come to life in a wonderful blend of geography, history, architecture, and archaeology. Children can participate in activities that highlight a different heritage: wearing native dress, bartering in the replicated marketplace, baking bread, learning traditional dances, and experimenting with their crafts.

Jewel/Osco

This mini-grocery-store exhibit, donated by Jewel/Osco, encourages children to learn while engaging in one of their favorite activities. Kids can fill their carts, weigh their produce, and use the cash registers—all of which are miniaturized. Kids are encouraged to touch and play with everything.

Where the Wild Things Are

This exhibit re-creates the journey of Max from Maurice Sendak's classic tale. Children can literally wander through life-size pages of the book, dress up in "Max" costumes, sail in his boat, or have a wild rumpus by the tent in the forest.

On the Road

Getting behind the wheel of a bright red "vintage" convertible, donning a helmet and trying a motorcycle, and driving in an actual simulator are adventures for kids to take off on in this exhibit.

The Learning Store

Tucked in the museum is the Learning Store, a hidden treasure that features an outstanding selection of toys. Among the loot are picture books, games, dolls, dinosaur coloring books, educational books, Lauri puzzles, and puppets.

Facilities and Access

Hours: Tuesday–Saturday, 10:00 A.M.–4:00 P.M.; Sunday, 12:00 noon–4:00 P.M.; Monday, closed. Learning Store: Tuesday–Friday, 9:00 A.M.–5:00 P.M.; Sunday, 12:00 noon–4:00 P.M.; Monday, closed.

Admission: $2.50 for adults and children, free for children under two. All children must be supervised by an adult age 18 or older.

Transportation and Parking: Going north on Lake Shore Drive, exit at Hollywood to Ridge, turning right on Ridge. Follow Ridge to Evanston, where it runs into Green Bay Road at Emerson. Stay left on Green Bay to 165 Green Bay Road in Wilmette. Parking is available in an adjacent lot or on Green Bay Road. Also, you may drop off passengers in front of the museum.

Restaurants and Food: There are no eating areas in the museum, but we gladly recommend Walker Brothers Pancake House, next door, at 153 Green Bay Road, Wilmette, (708) 251-6000.

Restrooms: Equipped with changing tables.

Information: Call (708) 256-6056. You can hear recorded information at (708) 251-7781. (Note: use the 312 area code before 11/11/89.)

MUSEUM OF BROADCAST COMMUNICATIONS

800 South Wells Street (at River City)
Chicago, Illinois 60607
(312) 987-1500

The Museum of Broadcast Communications bridges many important gaps: first, the gap between what entertains and educates your child and what entertains and educates you; second, the gap between television shows you watched as a child and the shows your children watch now; and third, the gaps between important elements of broadcasting—the historical, the hysterical, and the theatrical. Where else can you find—and actually watch—the Kennedy-Nixon debates, "Kukla, Fran and Ollie," and "The Shadow" alongside one another?

Tapes—Audio and Video

The museum encourages parents and children to share the fun of entertainment and education by viewing a tape from the extensive library of programs and commercials. When you enter the museum you will be greeted by some of old-time radio's favorite personalities, Charlie McCarthy and Mortimer Snerd. Enter the archives and pick out a tape to enjoy: tapes range from radio and television shows your children never heard of to local nightly news broadcasts dating from before your children were born to numerous dramas and educational programs, all of which you can watch in your own viewing and listening booth.

Exhibits

The museum regularly hosts special events that are open to the public. These include shows for children during holiday time and during the American Children's Television Festival. Recent attractions were visits from Bozo and Cooky, who were winners of the Ollie awards for kids' shows, and commemorative celebrations for anniversaries of children's shows both currently on the air and no longer broadcasting. These and other special showings are often scheduled in the 99-seat Kraft facility, also used for lectures and seminars. The theater is seldom dark and is almost always running something of interest—award-winning commercials, shorts, or shows.

Memorabilia

Before you leave the museum, be sure to look at the more traditional museum-like displays. They include a memorial to Pierre Andre in the form of a working radio console; dozens of old model televisions and radios, which, to a child, are practically unrecognizable as such; decoder rings from Little Orphan Annie; board games; books; and trinkets based on shows long gone.

Facilities and Access

Hours: 12:00 noon to 5:00 P.M. on Wednesday, Thursday, Friday, and Sunday; 10:00 A.M. to 5:00 P.M. on Saturday; and closed on Monday and Tuesday.

Admission: The suggested donation is $3.00 for adults, $2.00 for students, and $1.00 for seniors and children under 13.

Transportation and Parking: The museum is located two blocks south of the Eisenhower Expressway (Congress Parkway): exit 51-I off the Kennedy to Eisenhower east, exit 51-I off the Dan Ryan to Eisenhower east, or turn right at Wells Street at the end of the Eisenhower. Take the 22 Clark bus to Polk (it ends at Polk—the street just north of the museum), or take the 36 Broadway bus or the Lincoln Avenue bus to Harrison and Wells. For more information, call the CTA at (312) 836-7000. Parking is easily available in a lot just south of the building. The entrance as well as the entire museum are accessible to the handicapped.

Restaurants and Food: There is a deli downstairs in the building, a restaurant a few doors down on Polk, and outside seating along the river if you want to picnic.

Restrooms: These are located in the basement; no changing tables are available.

Information: Call (312) 987-1500.

MUSEUM OF THE CHICAGO ACADEMY OF SCIENCES

2001 North Clark Street (at Armitage Avenue)
Chicago, Illinois 60614
(312) 549-0606

The Museum of the Chicago Academy of Sciences opened its doors way back in 1857, but its exhibits are still exciting and educational for the late 20th-century kid. Focusing on the natural history of the Great Lakes region, the museum captures a child's attention with a lifelike atmosphere of grass, trees, and a great variety of wildlife figures—deer, owls, bears, foxes, and boars.

Children's Gallery

The first step at the Academy of Sciences, and the main attraction for kids, is the Children's Gallery on the third floor. This room at first appears small (capacity 20) and bland, but it comes alive when the museum staff brings out touchable snakes, frogs, and salamanders, and directs children toward puzzles, discovery boxes, and assorted educational games.

The gallery is designed for parent-child interaction, and

it is fun for all involved. Ask the staff for the following materials when you are in the Children's Gallery.

The Discovery Boxes are fantastic and a must for older children. There are about six boxes ranging in subjects from clams to chocolate to footprints. Parents will be surprised when they themselves learn something from opening a Discovery Box. Did you know the Aztecs used chocolate for money and to pay taxes? Can you identify the footprints of a snapping turtle? A beaver? A raccoon? These boxes can teach you something, too.

The Hand Puppets (or "Hanimals") are something all children enjoy. They are exotic and imaginative and range from octopi to owls. Older children often like to put together impromptu shows.

The Critters Board is where children put on masks of animals, place their heads in cardboard-cutout holes, and with the use of a mirror and forest-like backdrops, envision themselves as wild animals in the jungle. Don't be too bashful to try it yourself: it's fun.

Books are enjoyable for older and younger kids alike. The children's gallery provides a reading corner with special animal books.

Fossils and butterflies are available to handle; the children love them.

Fred the Snake is the star of the gallery (he can be touched), but there are also some popular frogs, birds, tadpoles, and salamanders.

The Children's Gallery is a special place. It contains great games, helpful staff, playful children's music, and numerous educational activities that parents and children can enjoy together. Be sure to call (312) 549-0775 (the Education Department) before you go, however, so you can check on the gallery's hours (which vary) and ask how crowded the room is likely to be upon your arrival.

Celestial Sphere and Conservation Theater

Once you've toured the Children's Gallery, visit the Celestial Sphere located on the same floor. The oldest planetarium in Chicago gives children a new understanding of the stars and constellations.

If the gallery and the sphere have exhausted some of your energy, take a breather in the Conservation Theater also on the third floor. Seven different videos (on dinosaurs!) are available to view.

Second Floor

The second floor is the central exhibit section of the museum. Dunes, prairies, woodlands, and marshes come alive with numerous plants and lifelike settings. Children can look eye-to-eye with realistic tigers, buckeyes, elms, and owls.

The museum creates an excitement by making you feel like you are outside in the wilderness. The exhibits, designed with animals and plants low to the ground, have a certain tranquility about them, and the waterfall creates an especially peaceful atmosphere.

On weekends the staff brings out a touch cart that captivates children. Kids are allowed to touch and grasp onto claws, shells, feathers, animal skins, and other artifacts.

In addition to all the interesting permanent exhibits, the museum has special, temporary exhibits that are often child-oriented. In the past it has had a fantastic Dino-Rama exhibit and an exciting mummy exhibit.

Classes

If your child has a particular interest in the natural sciences, be sure to pick up the academy's calendar of programs. The museum offers interesting classes on topics like bugs and nature myths. Also on the calendar are story-telling events, popular overnight parties (kids actually sleep in an exhibit), and nature walks in the Lincoln Park area.

Facilities and Access

Hours: The museum is open from 10:00 A.M. to 5:00 P.M. daily. Call for the Children's Gallery's hours.

Admission: Adults, $1.00; seniors and children (ages 3–17), $.50; children under 3, free. The museum is free to everyone on Mondays.

Transportation and Parking: The 22 Clark, 36 Broadway, 73 Armitage, 151 Sheridan, and 156 LaSalle buses all stop at the museum entrance. By car, take Lake Shore Drive or the 90/94 Kennedy Expressway. Street parking in Lincoln Park is very difficult, but there are less-crowded metered lots to the south and east of the museum.

Restaurants and Food: The nearest eatery is the Lincoln Park Zoo's cafeteria, which is directly behind the museum in the red brick building. The cafeteria, though, is noisy and only serves the basics, like hot dogs, chips, and Coke.

Restrooms: The toilets are located in the basement, and are small enough for children. The large countertop sink provides adequate space for diaper changing.

Information: For special information and specific questions not relating to hours and events, call (312) 549-0606. For 24-hour information including the standard info, call (312) 871-2668.

ORIENTAL INSTITUTE MUSEUM

1155 East 58th Street
Chicago, Illinois 60637
(312) 702-9514

The Oriental Institute Museum is a showcase of the history, art, and archaeology of the ancient Near East—Egypt, Mesopotamia, Syria/Palestine, Persia, and Anatolia. The museum offers a wide range of worthwhile tours and workshops for children and art-project-related kits for teachers. But unless your children are fascinated with archaeology or have a school-related project, we wouldn't recommend a full-day outing to this museum: this venerable, dusty treasure trove is perhaps more suited to academics than to our television-age children.

Mini-Museums on Egypt or Mesopotamia are available for loan to teachers at $10 for two weeks. Each mini-museum contains 12 or 13 objects, with an information card for each one. Call the Museum Education Office at (312) 702-9507 to reserve a mini-museum.

A six-session sketching program is offered each spring for children ages 12–18 who have a serious interest in art. Previous art instruction is not required, but advance registration is.

Slide talk sets are available on a variety of art topics. The kits must be reserved in advance and picked up at the museum. They contain 50 to 75 slides, and can be rented for $10 per week. They would benefit a sophisticated art class more than a class full of little ones.

Free group tours can be arranged by reservation only, and 30-minute films and slide talks are also available with reservations. There is a small fee of $.50 per person for the films and slide talks. Elementary- and advanced-level teacher's kits are available, too.

Sug Museum Store

The Sug ("market"), which is open during museum hours but closes at 3:30 P.M., contains unique gifts, jewelry, crafts, books, cards, posters, and prints.

Facilities and Access

Hours: Tuesdays–Saturdays, 10:00 A.M.–4:00 P.M.; Sundays, 12:00 noon–4:00 P.M. Closed Mondays but open on holidays. Free film series on Sundays at 2:00 P.M.

Admission: Free.

Transportation and Parking: To reach the Oriental Institute by CTA bus from downtown, take the Jeffrey Express bus south on the Outer Drive to 55th Street. Transfer to the 55th Street bus, getting off at Woodlawn Avenue. Walk south on Woodlawn to 58th Street. By car, follow the same route, and look for parking once you turn south on Woodlawn. The museum is wheelchair accessible.

Restrooms: Restrooms are available in the basement, down a flight of marble steps.

Information: Call (312) 702-9514 or (312) 702-9507.

PEACE MUSEUM

430 West Erie Street
Chicago, Illinois 60610
(312) 440-1860

If we are to reach real peace in this world and if we are to carry on a real war against war, we shall have to begin with the children.

Mahatma Gandhi

Gandhi knew what he was talking about: our children will determine the future of world peace, so we need to educate them. Keep your eye on the Peace Museum of Chicago. Although the majority of their exhibitions are geared toward adults, several times a year they feature outstanding programs for children.

Last year an exhibition called Play Fair focused on activities to develop communication and conflict-resolution skills in youngsters. Imaginative hands-on games helped children understand the causes of conflict and war. The Peace Museum also presented parents with different ideas and suggestions for explaining to their children historical tragedies involving war, such as the Holocaust.

As an outgrowth of this very successful exhibition, the

museum developed *Child's Play,* a resource guide related to children and peace education. Included in the resource guide is a listing of recommended children's books, a list of alternative non-war toys and games, organizations involved in peace education programs, and companies that carry nonviolent toys and games. The museum sells these guides for $9.45.

Although you should plan your visit to the Peace Museum to be at a time when it is showing a children's exhibition (call [312] 440-1860 for program information), a small children's center in the museum is open year-round. The children's center contains materials for drawing; a neat cotton-stuffed glove; mailboxes to Hiroshima, to Kiev (Chicago's sister city), and to the President in Washington, D.C.; and a poem machine that creates poems of peace from the names of children (a great trick).

Facilities and Access

Hours: Tuesday–Sunday, 12:00 noon to 5:00 P.M., but open Thursday until 8:00 P.M.; Monday, closed.

Admission: $3.50 for adults, $1.00 for seniors and students.

Transportation and Parking: By car, take Lake Shore Drive to LaSalle and LaSalle to Erie. Street parking is usually available. By el, take the subway to Grand and State and walk northwest.

Restaurants and Food: There is a good deli down the block and a variety of restaurants within walking distance.

Restrooms: Available on the first floor. No changing table.

Information: Call (312) 440-1860.

TERRA MUSEUM OF AMERICAN ART AND MUSEUM OF CONTEMPORARY ART

When trying to think of an interesting weekend activity for children, most parents run through the traditional list of a ball game, zoo, park, or science museum. But, fortunately, a few wise parents have found an addition to the list that offers unique enjoyment for them and their children—the downtown art museums. Some people think that art museums are not able to capture the attention of young children (let alone many adults), but they are wrong. Surprisingly, children are often more adept at understanding artwork than many adults.

"Children are more open-minded and perceptive when looking at paintings," said a veteran docent at the Terra Museum. "They frequently teach and show me things that I had never considered nor seen in a work, and they almost always grasp the meanings of abstract pieces faster than adults."

The key to enjoying art museums with your children is to ask questions that draw them into reacting. How does a particular painting make them feel and think? What memories does the picture conjure up? You may want to seek assistance from one of the museum docents, who are particularly skilled at enticing children to respond to the art pieces.

Terra Museum of American Art

644 North Michigan Avenue
Chicago, Illinois 60611
(312) 664-3939

Founded only a few years ago by the famous entrepreneur and cultural ambassador Daniel J. Terra, the Terra Museum of American Art is one of a few museums in the United States devoted exclusively to American art. The nationally renowned masterpiece, *Gallery of the Louvre* by

Samuel Morse, and the museum's excellent collection of luminist and American impressionist works have established the Terra Museum's distinguished reputation.

When you plan to tour with children, we recommend calling the museum in advance at (312) 664-3939 to reserve a tour given by a docent and to see if there are any family activities on the schedule. Personal tours are the most exciting way to visit the museum (and are the least work for parents). Docents offer all sorts of activities, such as looking at paintings while lying flat on your back, that make the visit more enjoyable for both you and your children.

If you are unable to reserve a personal tour (minimum of six people required), you can still join the daily scheduled tours at noon and 2:00 P.M., but these tours are focused mostly toward adults and older children. The presence of children may elicit an alteration in the tour guide's style, however.

In the past, the museum has sponsored events in the galleries that were geared to and suited for children: arts and crafts, treasure hunts, and assorted games. Because of funding difficulties, though, the museum may curtail its family activities.

Another good option is to obtain from the front desk a self-guide pamphlet for parents and teachers. This guide is extremely helpful in directing parents to artwork that children like, and it offers good questions and subjects that promote discussion with the children.

Two incidental activities, unrelated to the artwork, are frequent child-pleasers at the Terra Museum: riding the gigantic elevators that haul large works of art (along with visitors) upstairs, and looking out the upstairs window at the wonderful view of Michigan Avenue. Many children have never seen Michigan Avenue from that perspective and virtually no child and few adults have ever been in an elevator like the one in the Terra Museum (it's as big as a large room).

Museum of Contemporary Art
237 East Ontario Street
Chicago, Illinois 60611
(312) 280-2660

The MCA, as it's popularly called, offers a rotating series of exhibitions that are usually more complex, innovative, and experimental than the work shown at the Terra Museum. The exhibitions, depending on their nature, may or may not be enjoyable for children. Because the art is contemporary, children often relate to the work easily. Sometimes the exhibits are just too sophisticated or too wacky for kids—and for many adults. The MCA is only a few blocks away from the Terra Museum, however, and visiting the MCA after the Terra Museum makes for an interesting contrast.

When you arrive, ask at the front desk if the current exhibit is one children would enjoy. If they say no, then head downstairs to the pleasant cafe, have something to eat, and show your children the two exotic and amusing sculptures in the basement. If you and the MCA staff think the exhibit may be good for children, then try to join a free tour (12:15 P.M. on weekdays and 1:00 P.M. and 3:00 P.M. on weekends). If a tour is not available, just wander the interesting museum on your own.

Facilities and Access

Hours: The Terra Museum is open Tuesday, 12:00 noon–8:00 P.M.; Wednesday–Saturday, 10:00 A.M.–5:00 P.M.; and Sunday, 12:00 noon–5:00 P.M. It is closed Monday.

The MCA is open Tuesday–Saturday, 10:00 A.M.–5:00 P.M.; and Sunday, 12:00 noon–5:00 P.M. It is also closed Monday.

Admission: The Terra Museum charges $4.00 for adults; $2.50 for seniors; and $1.00 for students with IDs. Children under 12 are admitted for free. Admission to the MCA is discretionary.

Transportation and Parking: Both museums may be reached by Lake Shore Drive to Michigan Avenue and by CTA buses running along Michigan Avenue. The nearest subway station is at Chicago and State. Metered parking is extremely difficult downtown. There are various nearby lots charging steep hourly fees.

Restaurants and Food: There is no restaurant in the Terra Museum. The MCA has a cafe in the basement with good snacks. Michigan Avenue restaurants provide a variety of choices, from fast food to fancy dining.

Restrooms: Child-accessible bathrooms are available in both museums; however, there are no changing tables.

Information: Call the Terra Museum at (312) 664-3939 and the MCA recording at (312) 280-5161.

4
CHICAGO'S ETHNIC MUSEUMS

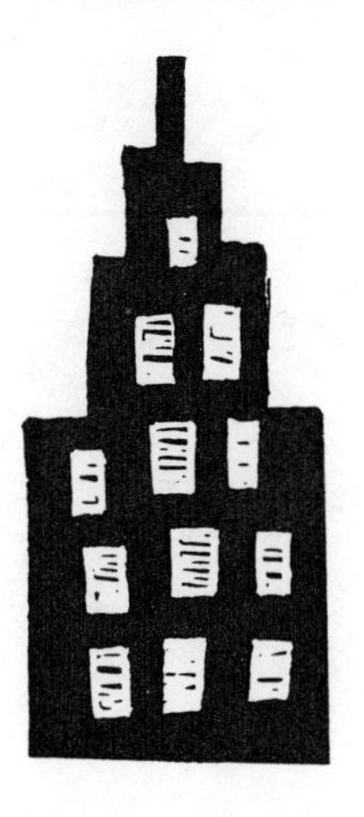

BALZEKAS MUSEUM OF LITHUANIAN CULTURE

6500 South Pulaski Road
Chicago, Illinois 60629
(312) 582-6500

If it's the second day of summer and your children are telling you for the one-thousandth time, "There's nothing to do," pull out a globe and ask them to find Lithuania. After a short moment of frustration (or a long moment depending on your mood), inform them that the country no longer exists but that you can magically take them there for a short time, if they want to go (Lithuania was annexed by the USSR in 1940). Then, with their curiosity aroused, quickly whisk them away for the day to the Balzekas Museum.

The museum provides children with a unique cultural perspective on an Eastern European people, and because of the excellent children's gallery on the first floor of the museum, the visit is a great deal of fun. The newly expanded museum features a half-hour video show (in the first floor audio-visual room) on Lithuanian culture and traditions; historical documentation of this Baltic people's experiences; and exhibits on native costumes, folk art, amber, decorative objects, and antique armor and weapons.

Children's Gallery

The Children's Gallery is divided into two sections: the 19th-century period and the medieval times, which are appropriately separated in the middle by a little drawbridge. The medieval section is the real child-pleaser, since it contains a coat-of-arms activity—a Castle Closet, where kids dress up in costumes—and a four-foot wooden puzzle of a man in armor. Kids can play different, chivalrous games with the various toys in the gallery and can imitate the 13th-century Lithuanian fight against the Teutonic knights. Kids love to place their heads in a cardboard-cutout medieval scene in this room, pretending to be old knights in shining armor while their parents photograph them in their glory.

The 19th-century scene is a re-creation of an old farmstead and a depiction of rural life. A large model of a thatched farmhouse is in the center of the room and lining the sides are various Lithuanian ornaments, like beautiful dolls and a Kaufles instrument that looks like an old-fashioned harp. Usually, a staff member is available to aid in the supervision of games and activities. On several weekends throughout the year, there are arts and crafts festivals that are even more fun for children. It is a good idea to call Pat Bakunas at (312) 582-6500 to find out the museum's schedule of events.

The Balzekas Museum has a 22,000-volume library covering the range of historical and cultural Lithuanian topics. If your child ever has to write an essay or report on Lithuania, this is the place to come. Next to the library is the Women's Guild Room, which exhibits extraordinary, handwoven dresses made more than a hundred years ago, and beautifully crafted dolls, necklaces, and shoes.

The museum also has a Holocaust room, an old-map gallery, and numerous Lithuanian artifacts, perhaps of greater interest to older children and adults than to younger children. An excellent gift shop is located right next to the entrance, containing Lithuanian dolls, wood carvings, crystal, and more items crafted in Lithuania or in its style.

Facilities and Access

Hours: The museum is open Saturday–Thursday from 10:00 A.M. to 4:00 P.M. and Friday from 10:00 A.M. to 8:00 P.M.

Admission: $1.00 for children under 12; $2.00 for students and seniors; and $3.00 for adults.

Transportation and Parking: The Stevenson Expressway crosses Pulaski about five miles north of the museum. Pulaski Avenue CTA buses also run by the museum.

Restaurants and Food: There is no food available in the museum. However, recommended favorites for authentic Lithuanian food are Tulip Restaurant, 2447 W. 69th Street, (312) 925-1123, and Healthy Food Lithuanian Restaurant at 3236 South Halsted Street in Bridgeport, (312) 326-2724.

Restrooms: Child-accessible bathrooms are located on the first floor, but there is no changing table available.

Information: Call (312) 582-6500.

DUSABLE MUSEUM OF AFRICAN-AMERICAN HISTORY

740 East 56th Place
Chicago, Illinois 60637
(312) 947-0600

Way back in 1779, long before the Sears Tower, Water Tower Place, or any other western settlement here, Jean

Baptiste DuSable erected a log cabin on the north bank of the Chicago River, which made him the first permanent settler in this city. If your children don't know about black heroes like DuSable, Frederick Douglass, Marcus Garvey, and Malcolm X, then visit the museum that bears DuSable's name: the DuSable Museum of African-American History.

The DuSable Museum does not focus on youngsters, yet the current, topical nature of many of its exhibits lends itself to parent-child interaction. Looking at historic pictures of Angela Davis, Julian Bond, Jesse Jackson, or the Sixties Freedom Struggle in Selma provides an opportunity for parents to expound upon what they or their families were feeling, thinking, and doing during the Civil Rights Movement.

In addition to the many recent, evocative photographs, the museum also devotes considerable space to a celebration of African culture. Volunteer museum guides lead children through the musical, artistic, and religious displays, highlighting points of interest. Many of these guides (or docents) are retired teachers, and they engage children with questions such as, "Tell me which of the African masks are sad and which are happy? Why do you think some of the masks have horns?"

If you can assemble a group of 10 or more people, the museum also offers a special children's tour. This tour consists of a brief lecture, a short film, a hands-on exploration of African musical instruments, and a guided tour through selected galleries. Movies shown in the past have included African folk tales, such as *The Magic Tree* and *An-an-si the Spider.*

The museum conducts special workshops for children. Some past themes of these workshops have been mask making, costume making, the Kwanza celebration, the island of Hispaniola, and Haitian carnival. During the fall and winter months, DuSable sponsors a minority-oriented book fair and a children's film festival. For more information, call the museum and request their most recent calendar of events.

Below is a description of DuSable's more interesting attractions for children.

Slave Gallery

This is a small room depicting the life of a slave. A diagram of a slave ship explains that each passenger had as much room as there is space in a coffin. On the floor and walls are shackles, old-fashioned pressing combs, scrub boards, and pails.

Generations in Struggle

A time line rings this room, dating from 1700 to the present. The time line depicts the black struggle in America from the slave rebellion of Nat Turner to the election of Chicago's first black mayor, Harold Washington. A small, free-standing exhibit on segregation is educational for children. There is also a replica of a typical Mississippi diner, complete with ketchup and mustard. Above the countertop a sign reads, "Whites Only, Colored Use the Back Door."

Treasures of the DuSable Museum

This room contains artifacts and representations of American black culture. Joe Louis's boxing gloves are fun to look at, as well as Paul Laurence Dunbar's marble writing table.

Gift Shop

Several children's items are available at the gift shop: black science coloring books, a map of Africa, kise stone animals, and a black history series.

Membership

DuSable offers a variety of membership categories, ranging from a senior citizen membership ($15 annually) to the annual sustainer ($1,000). Membership entitles you to a free Heritage calendar, the newsletter, invitations to special events, etc. Call and request the membership brochure.

Expansion

Sometime in 1989, DuSable will undergo an expansion of 75,000 square feet, more than twice the size of the existing museum. Included in the planned expansion is a 500-seat theater and a full-service cafeteria. The museum hopes to use some of the additional space for classes on African art and dance.

Facilities and Access

Hours: Monday–Friday, 9:00 A.M.–5:00 P.M.; Saturday and Sunday, 12:00 noon–5:00 P.M. Closed New Year's Day, Easter, Thanksgiving, and Christmas.

Admission: $2.00 for adults; $1.00 for seniors and students; and $.50 for children under 13. Free on Thursday.

Transportation and Parking: Located in an uncongested area a good distance from downtown, the museum is most easily accessible by car. Behind the building there is a large parking lot. To arrive by public transportation take the 4 Cottage Grove bus and get off at 57th Street. Or take the el to 55th Street and transfer to the west-bound 55th Street bus.

Restaurants and Food: The current building does not have an eating facility, although a full-service cafeteria is planned for 1989. Museum staffers recommend two local restaurants: Army and Lou's at 422 East 75th Street, (312) 483-6550, and Gladys' Luncheonette at 4527 South Indiana Avenue, (312) 548-4566.

Restrooms: On the same level as museum; changing table in women's restroom only.

Information: Call (312) 947-0600.

LATVIAN FOLK ART MUSEUM

4146 North Elston Avenue
(entrance on Hamlin Avenue)
Chicago, Illinois 60618
(312) 588-2085

Just as Latvia is nestled between the Baltic Sea on the west, Russia on the east, Estonia on the north, and Lithuania on the south, this museum is nestled in a northern corner of the city, tucked away on the second floor of an inconspicuous building. The museum is in the Latvian Community Center, which serves Chicagoland's 5,000 Latvians. This culture is being threatened by the occupying government in Latvia today, so it is increasingly critical for us to learn how Latvian traditions can thrive. Among the regions represented at the museum are the Riga, Zemgale, Letgale, and Kurzeme. All the descriptions are bilingual, in both English and the Baltic branch of Indo-European languages spoken·in Latvia. The collection is rich in textiles, woven items, and ceremonial dress clothing.

The best way to approach the museum is to use as a guide the descriptive catalog for sale at the counter. It describes the wide array of coins, jewelry, fabrics, dolls, and archaeological digs. The bronze, silver, and amber jewelry displayed are from as early as the 11th century. There are lovely ceramic vases and jars, ornamental needlework, and practical spinning wheels and distaffs. Upon request the museum staff will uncase the Latvian musical instruments, such as the kokle, for children to touch or play. Traditional folk costumes, brooches, and hats line the walls of the museum. The excellent catalog maps out the geographic and chronological origins of every object in the museum. The staff is very willing to answer questions, is tolerant of children, and offers information and background on the history of Latvia and the Latvian community in the Chicago area.

Facilities and Access

Hours: By appointment only. Call (312) 588-2085, the Latvian Community Center, for reservations.

Admission: Free, but donations are appreciated.

Transportation and Parking: Take Lake Shore Drive to Montrose, Montrose west to Elston Avenue, turn right (north) onto Elston. Metered parking is available on the street. The museum entrance is on Hamlin.

Restaurants and Food: There is no food service in the museum. Call and ask the community center personnel for their recommendations.

Restrooms: Located on the first floor.

Information: The museum itself has no phone, but you may reach the Latvian Community Center office at (312) 588-2085.

MEXICAN FINE ARTS MUSEUM

1852 West 19th Street
Chicago, Illinois 60603
(312) 738-1503

When your children's heads are filled with frightening thoughts of goblins, monsters, and ghosts around Halloween time, try something different and stop by the Mexican Fine Arts Museum to check out the Day of the Dead Exhibit (*El Dia De Los Muertos*). Your children will be in for a delirious Halloween shock. The Mexican Halloween is vastly different from the Halloween of the *Norteamericanos.* Instead of touring the neighborhood in ghastly garb and asking for candy, many Mexicans celebrate Halloween with a fiesta at their family's graveyard. Traditionally, the entire living family joins in celebration with those family members who have died. Little ones frisk amid the crucifixes and tombstones, playing games and eating treats. Older children may set up a game of tiny dice on top of their grandparents' tombs. The youngest child might arrange miniature wooden and cardboard furniture, setting the little table with pea-sized pottery dishes. And everyone celebrates, singing songs around the graves of the dead.

The Day of the Dead exhibit, which is open from late October to early December, is the most popular visiting time for children at the Mexican Fine Arts Museum. Many of the art exhibits here are typically more adult-focused. During the celebration period of this intriguing Mexican tradition, however, there are activities, skulls, alters, paintings, and photographs that will captivate children's attention.

Christmas at the Mexican Fine Arts Museum is similarly child-oriented. The staff leads a game called the Three Wise Men and plans other celebratory Christmas activities to compliment the artworks of the gallery. For both the Halloween and Christmas exhibits, call (312) 738-1503 to schedule your visit and make a reservation for a tour to heighten your children's ability to appreciate the center.

The Mexican Fine Arts Center is more of an art gallery than a museum. Like with the Terra Museum of Art and other Chicago art museums, we have found that kids can appreciate the art when asked questions that draw them into it. The center schedules special exhibits throughout the year: some excellent exhibits in the past have included Living Maya, Barrio Murals, Latino Art, Mexican Master Weavers, and Images of Faith.

The center also provides free fine-art and dramatic-art classes. Topics covered recently are theater (children ages 9–14), photography (for all ages), and drawing for youth and adults. Reservations are required for all classes, so call or write the Mexican Fine Arts Center Museum.

Beautiful prints, native crafts, and postcards are available at somewhat steep prices at the front office of the center.

Facilities and Access

Hours: The museum is open Tuesday through Sunday from 10:00 A.M. to 5:00 P.M. It's closed on Mondays.

Admission: Free. Contributions welcome.

Transportation and Parking: From downtown, take the Dan Ryan south to 19th Street and take a right. Or the Eisenhower or Stevenson will put you on Damen Avenue, which crosses 19th Street two blocks from the center. Parking is available on the street in front.

Restaurants and Food: The museum offers no food facilities. However, Nuevo Leon Restaurant, 1515 West 18th Street, (312) 421-1517, is three blocks east of the center in the heart of the Pilsen community. It serves fantastic Mexican food and is very reasonably priced and very kid-friendly. Also Parrillita Restaurant, 1409 West 18th Street, (312) 421-9779, is a good alternative.

Restrooms: The restroom is near the office and accessible to children, but there are no changing tables.

Information: Call (312) 738-1503.

POLISH MUSEUM OF AMERICA

984 North Milwaukee Avenue
Chicago, Illinois 60622
(312) 384-3352

I came here, where freedom is being defended, to serve it, to live or die for it.

General Kazimierz Pulaski to his commanding officer, General George Washington, 1777.

And you thought Pulaski was only a long street flanking the west side of Chicago. Like the Italian-American patriarch, Giuseppe Garibaldi, General Kazimierz Pulaski came to America from his homeland in 1776 to fight alongside the colonists. Expulsed from Poland for resisting Russian tyranny, Pulaski worked tirelessly to train the American troops in modern European warfare, eventually earning himself the title "Father of the American Calvary." He ultimately gave his life for the American Revolution, dying at the Battle of Savannah in 1779.

Bring your children to the Polish Museum of America and introduce them to a long line of Polish and Polish-American people of extraordinary accomplishment. General Pulaski is one Polish great. Copernicus, the charter of the skies and founder of modern astronomy, is another. Why, even the discoverer of radium and first woman to win the Nobel Prize is of Polish descent—Maria Sklodowska-Curie.

The Polish Museum of America can be a great place to visit for children, but only if you're willing to put in some extra effort explaining the exhibits and fielding questions: it's not Disneyworld. Most of the displays sit inside of glass cases, and there are no activities or tours designed solely for kids (except during the Christmas season; see Children's Program below). If you're good at talking things over, however, the exhibits will spark your child's curiosity.

If your only knowledge of Polish history is Roman Pucinski, Richie Zisk, and Lech Walesa, don't panic. The museum staff is very dedicated, very helpful, and very Polish. The many Polish visitors are usually very knowledgeable, and they can provide living contact with Polish culture for your child. In fact, Chicago is filled with Polish culture, for it has the largest Polish community outside of Warsaw. Below is a listing of some of the highlights in the museum.

Main Viewing Room

Children gravitate toward the two more colorful and eye-catching exhibits in this room—the mannequins dressed in traditional costumes and the rows upon rows of painted Easter eggs. Displays honoring dozens of Polish heroes line the walls of this room.

Boat Room

If your son or daughter builds model airplanes or ships, then this small display is a must. This room features dozens of miniature, yet real-looking Polish ships, including the *Patory*).

Paderewski Room

The memorabilia in this room, entirely dedicated to the memory of Polish statesman and musician Ignacy Jan Paderewski (1860–1941), include the famous fur coat, letters in his own hand, the last piano he ever touched, and even the purple sofa that graced his living room. The personal nature of this exhibit is intriguing to children and adults alike. It sticks with the ordinary paraphernalia of everyday life that even people of achievement possess.

Gift Shop

A visual delight for children, this shop offers a variety of handcrafted items from Poland—Fujarki flutes; wooden puppets; and dolls dressed in traditional, regional costumes.

Children's Program

During the three weekends preceding Christmas, the Polish Museum sponsors traditional music and dance shows for children. This mini-festival celebrates the St. Nicholas holiday (December 6) and usually involves choir singing, choral groups, and gift giving. The public is welcome, but call first for scheduling details.

Facilities and Access

Hours: Everyday, 12:00 noon to 5:00 P.M. Closed Good Friday and Christmas.

Admission: Free.

Transportation and Parking: The easiest way to get to the Polish Museum is by car. The building is located right off the Kennedy Expressway at Milwaukee and Augusta. To arrive there by bus, take the Milwaukee Avenue bus. The closest subway station is the Chicago Avenue stop on the O'Hare line, about six blocks from the museum.

Restaurants and Food: While there is no place to eat within the building, there are two nearby Polish restaurants, which come highly recommended: Mareva's at 1250 N. Milwaukee Avenue, (312) 227-4000, and Podhalanka at 1549 W. Division, (312) 486-6655. Mareva's is the fancier of the two and Podhalanka is less expensive.

Restrooms: There are restrooms on the second floor, but neither the men's nor women's room has a changing table.

Information: Call (312) 384-3352, ask for the curator, Jack Nowakowski (pronounced Novakowski).

Polish Museum Library: Open on weekdays: Monday and Friday, 12:30 P.M.–7:30 P.M.; Tuesday–Thursday, 10:00 A.M.–6:00 P.M. Adjacent to the museum, this library has over 30,000 volumes relating to Poland.

SPERTUS MUSEUM

618 South Michigan Avenue
Chicago, Illinois 60605
(312) 922-9012

If genocide is to be eliminated, we must understand what happened in the past.

Those are the first words visitors read when they enter the Bernard and Rochelle Zell Holocaust Memorial. Those words along with the collection of shocking Holocaust photographs and artifacts—a concentration camp uniform and rings torn from the hands of victims—remain in the minds of anyone who enters the Spertus Museum.

The Spertus Museum does not focus on the Holocaust alone, though, and a visit can be an opportunity to gain an understanding of Jewish culture and traditions. But when you are with young children, the Holocaust is hard to ignore. Questions in your mind inevitably pop up, such as how can you explain such a horrible, recent event in human history to children without filling them with nightmarish thoughts? And should you even expose your children to such facts now? These are good questions, and they are questions you should be prepared to answer before you enter the Spertus Museum.

If older children (and adults) are prepared for it, the Holocaust Memorial can provide them with a larger knowledge of the world and a wider understanding of human ignorance and our capacity for evil. For those parents interested in exposing their children to the Holocaust but unsure of how to go about it, a helpful parable is found on the memorial wall near the exit. It says, "The Holocaust centers around a basic paradox: It imposes silence, but demands speech. It defies solutions, but requires response."

Permanent Collection

The central gallery of the museum focuses on the core of Jewish traditions and the various rites of passage in Jewish culture. There are displays covering events from marriage and the bar mitzvah to the Sabbath and the Day of Atonement. Old and wonderful ceremonial objects, such as a 17th-century Hanukkah lamp and an 18th-century pewter laver used for ablutions, are included in the displays. Without a guide, however, it is difficult to bring the museum to life for children. Call (312) 922-9012 in advance to arrange a tour by a museum docent and to inquire if there are any upcoming, special programs for children. Occasionally, the museum has hands-on workshops for children where they can make their own Torah binders or mezzuzahs. There are also some programs including mime and storytelling.

The Spertus Museum is well known for containing works of Philip Pearlstein, Abraham Ratner, Andy Warhol, Friedrich Adler, and other world-famous, modern artists along with a vast array of costumes, jewelry, coins, artifacts, sculptures, and textiles. The pieces in the collection represent the far-flung history of 3,500 years of

Jewish migration and settlement, and they make this one of Chicago's more intriguing ethnic museums.

The museum maintains an excellent schedule of special exhibitions that may or may not be child-oriented. In the past, subjects have ranged from Yiddish Theater to the trial of Adolf Eichmann. Again, call in advance to find out what is currently showing.

Facilities and Access

Hours: The main gallery is open Sunday–Thursday, 10:00 A.M. to 5:00 P.M. and Friday, 10:00 A.M. to 3:00 P.M. It's closed on Saturday. The artifacts center (children's museum) is open Sunday–Thursday, 1:30 P.M.–4:30 P.M. Closed Friday and Saturday.

Admission: $3.50 for adults and $2.00 for children over two and seniors.

Transportation and Parking: The museum is in the South Loop and accessible by Lake Shore Drive or the el (Jackson Street stop). Metered parking is difficult to find, but there are various nearby garages in the area that charge reasonable rates.

Restaurants and Food: Walk down Michigan Avenue and you will have your choice of various eating spots.

Restrooms: There are restrooms on both the first and second floors that are handicapped accessible and will accommodate a stroller.

Information: Call (312) 922-9012.

SWEDISH AMERICAN MUSEUM CENTER

5211 North Clark Street
Chicago, Illinois 60640
(312) 728-8111

The comfortable 24,000-square-foot Museum Center is a relaxing and interesting visit. See some old and beautiful handcrafted Swedish weavings, embroidery, and woodwork. Check out the model depictions of early Swedish settlers in Chicago and the artifacts of their daily lives. There are antique waffle irons, a mini-distillery, and funny-looking scales (200 years old). A beautiful wall mural depicting a logging operation in "old" Sweden dominates the central gallery.

On a regular day, however, the museum's main gallery will not hold a child's attention for very long. If you call Kirsten Lane at (312) 728-8111 before your visit, though, she will tell you about the special schedule of events at the museum, which have included songfests, harvest-time celebrations, basket making, and Swedish fairy tales.

Special programs and tours can be arranged if you call in advance.

Facilities and Access

Hours: The Swedish Museum is open 11:00 A.M.–4:00 P.M. on Tuesday through Friday and 11:00 A.M.–3:00 P.M. on Saturday and Sunday.

Admission: $1.00 for adults and $.50 for children.

Transportation and Parking: The museum is located about one hundred feet north of the corner of Foster and Clark. Taking Lake Shore Drive to Foster or just driving down Clark Street are the quickest car routes. The 22 Clark bus stops near the museum. There is metered and free parking on Clark Street and Foster Avenue.

Restaurants and Food: After visiting the museum, a stop at Ann Sather's Restaurant, two doors south, is a must. The homestyle Swedish cooking and pleasant atmosphere will make you wish you lived in Sweden. Try the tasty cinnamon rolls or the famous pies if you only have time for a snack. If you have more time, though, the balance of the menu is excellent and the style of cooking is simple, wholesome, and straightforward—the kind your children will surely enjoy.

Restrooms: There are no changing tables, but child-accessible bathrooms are located in the back of the main gallery.

Information: Call (312) 728-8111.

UKRAINIAN NATIONAL MUSEUM

2453 West Chicago Avenue
Chicago, Illinois 60622
(312) 276-6565

The Ukrainian National Museum is more of a home than a museum. Except for a small plaque on the front door, the building looks like a modest, residential house. When you call about a visit, your inquiries about "official hours" and "special children's programs" will be greeted with a gracious, "Come when you want. I'll wait for you. And when you arrive, tell me what your children would like to see." (For the record, the "official hours" of the museum are from 12:00 noon to 3:00 P.M. on Sunday.) Once inside, you'll probably wait for your guide for 5 to 10 minutes as he or she banters back and forth in Russian with the other Ukrainians present. Be prepared to chitchat, because your guide will also probably introduce you to whoever is there.

This type of personal, expert attention sets the Ukrainian National Museum apart from other museums in the city. (Imagine scientists showing you and your children around the Museum of Science and Industry, or artists chatting and strolling with you through the Art Institute!) At the Ukrainian Museum you will find a dedicated and knowledgeable Ukrainian who is eager to share stories with your children; enlighten them about Ukrainian customs, art, and history; and, perhaps most importantly, provide a living example of fierce Ukrainian pride and sense of national identity, despite the fact that Ukraine is part of the Soviet Union. Below is a description of some of the highlights in the museum, as well as memorable quotes and stories from one of the tour guides, Olha.

Ukrainian History Exhibit

Olha spent over 20 minutes discussing a map of Central Europe and explaining the origin of the Ukrainian national identity and her people's plight in the last 1,000 years. Curious children will have lots of questions and will help set the course of your visit. "Ukrainians are so hospitable and so attractive as a culture," said Olha, "that the Vikings halted their southward expansion in the Ukraine, many of their warriors marrying Ukrainian girls and entering our culture."

Ukrainian Pottery and Weaving

Pointing to the difference between two sets of vases and teacups, Olha explained that when someone asked a father for his daughter's hand in marriage, the father's answer was communicated by his choice of tea sets. If the father chose the elegant set, his answer was yes, and the celebration began. If his answer was no, he chose the more mundane pottery. In this way, the father communicated his feelings without saying a word. "Sometimes symbols are better than words, don't you think?" asked Olha. Stories and legends lie behind all of the displays. Ask Olha—if she happens to be your guide—about the hand-carved chess set made by her father.

Ukrainian Colored Eggs

"Before people even knew if the sun revolved around the earth or the earth around the sun," said Olha, "Ukrainians knew that the sun created life. For this reason, they believed that the yolk inside of the egg symbolized the sun and represented life. That is why we paint eggs at Easter time." Your children will love these brightly colored eggs. Ask your guide for the sheet explaining the various symbols (for instance, a deer means wealth and prosperity, a fir tree means eternal youth, and a flower means love and charity).

Ukrainian-American Photo Gallery

This wonderful set of photos, like the one of the 1912 Ukrainian Choir of Chicago, honors the achievements of the 1.5 million first-, second-, and third-generation Ukrainians who live in the United States (over 60,000 in Chicago).

Facilities and Access

Hours: 12:00 noon to 3:00 P.M. Sundays. All other days, by appointment only.

Admission: Olha said, "The Ukrainians have a saying, 'If you don't like it, you don't pay.' So don't pay me until after the visit, and only pay me if you enjoy it." Posted admission: $1.00 for adults, $.50 for students.

Transportation and Parking: Located in an uncongested part of the city, the museum offers plenty of parking. The O'Hare el stops six blocks away (the Chicago Avenue stop). The Chicago Avenue bus also runs past the museum's front door.

Restaurants and Food: The museum is located in a neighborhood called the "Ukrainian Village," so there are numerous Ukrainian restaurants nearby. Olha recommended Galan's at 2210 West Chicago, (312) 292-1000.

Restrooms: The restrooms do not have special changing tables, but the staff does not mind your changing your baby on the floor.

Information: Call (312) 276-6565.

5
PARKS, ZOOS, AND OUTDOOR EXCURSIONS

BOTANICAL GARDENS

The Chicago area has several excellent botanical gardens in the suburban areas, each of which would be a pleasant outing for a garden- or nature-conscious family. The two most interesting are the Chicago Botanic Garden and Morton Arboretum. Each is approximately a 45-minute drive from downtown and each has special features and programs that have the makings of a great family outing.

The Chicago Botanic Garden, adjacent to the Skokie Lagoons on Lake Cook Road in Glencoe (a half mile east of Edens Expressway, [708] 835-5440) hosts innumerable garden shows on its very busy calendar. It also has a Japanese garden that distills the essence of peaceful space and meticulous cultivation in sharp contrast with the reality of both Japan's and our crowded urban lifestyle.

Morton Arboretum, on Route 53 in Lyle, Illinois, (708) 719-2400, north of the East-West Tollway Route 88, is an even more expansive site (1,500 acres) and is especially impressive during the spring blossom time and the fall change of colors. Call ahead to see if their tea garden snack shop is open and to find out what special activities are scheduled for the times you want to visit. Admission is $3 per car. (Note: use the 312 area code before 11/11/89.)

BROOKFIELD ZOO
8400 West 31st Street
Brookfield, Illinois 60513
(312) 242-2630 (Chicago phone number)

Brookfield Zoo is a vast delight of a zoo, one of the first to experiment with natural habitats for animals. But a trip to Brookfield can be an exhausting ordeal with small children if you hit the gates at the same time as the summer crowds. Instead, consider visiting the zoo when many others may not—those days when your children may have an unusual school holiday, or in winter, when the zoo is lightly visited and makes you feel like you are in a foreign land, with just you and beasts of the wild. On the other hand, if you are up to it, brave the crowds and visit during some of Brookfield's many holiday celebrations—they are a lot of fun.

For the past several Decembers, the Brookfield Zoo has had a Holiday Magic Festival, during which the zoo is open winter weekend evening hours before Christmas, when the less hardy might not venture out. Not only does the zoo decorate its 200-plus acres with twinkling lights, but it also has local schools and community groups decorate its trees with handmade, animal-safe ornaments. Costumed human-size animals, Frosty the Snowman, craft shop elves, mimes, musicians, storytellers, jugglers, and magicians all roam the grounds during this festival, and the Snowball Express bus provides free (and heated) transportation around the grounds during these evening hours.

Brookfield Zoo celebrates all holidays and makes up a few of its own during the year—including a teddy bear birthday picnic, Halloween costume parades, and a Ground Hog Day celebration complete with media breathlessly awaiting the pleasure of Mr. Groundhog.

In addition, the zoo schedules numerous seminars for children—including Backstage at the Zoo—in many of its habitats (Australia House, Tropic World, the Aquatic and Perching Bird House, among others). The Nature's Neighbors seminar helps children notice the wildlife that live within the city, and there's a session to meet the zoo's veterinarian. Most of these events have a modest charge, but they guarantee a place for you and your children without a crowd to prevent your enjoyment of this splendid zoo.

Plan to spend a full day in order to see all the exhibits and special programs. Set in 204 acres of landscaped scenery, Brookfield is a perfect day's outing. Bring umbrellas for rain and a big blanket in case you want a midday nap between all the walking! And don't forget to get a map of the 204-acre zoo, or you will get lost.

Special Events

Dolphin Shows: Adults, $2.00; children, $1.50; 65 and over, $1.50; under 3, free. Summer schedule: weekends and Tuesdays at 11:30 A.M., 1:00 P.M., 2:30 P.M., and 4:00 P.M. All other days, 11:30 A.M., 2:30 P.M., and 4:00 P.M. Hours change for fall and winter. This show is much less glitzy and much more informative than several of the commercial operations.

Tropic World: Opens at 10:30 A.M. and closes one hour before zoo closing. Free with zoo admission. The world's largest exhibit of its kind, this lavish depiction of the rain forests of Asia, Africa, and South America is home to a wide array of primates, other tropical mammals, and exotic birds. And it really rains inside Tropic World on a regular schedule so the animals will feel at home!

Children's Zoo: Adults, $1.00; children, $.50; 65 and over, $.50; youngsters under 3, free. Special shows include cow/goat milking and an animals show. This is a petting zoo.

Elephant Demonstration: Memorial Day to Labor Day only. Wednesday and Thursday at 2:30 P.M.; Tuesday

Friday, weekends, and holidays at 11:30 A.M. and 2:30 P.M. The vast size and gentleness of the elephants are very impressive to children, and sometimes the staff even lets the kids touch the elephants.

Feeding Schedule: Bear Grottos 3:00 P.M. and Aquatic Bird House 4:00 P.M. For some reason, people (especially children) like to watch animals eat. Watching an owl chew a frozen mouse grasped tightly in its talons is a treat for some children and can be a real highlight.

Motor Safari: Adults, $1.75; children ages 3–11, $.75; 65 and over, $.75; youngsters under 3, free. This 45-minute guided tour over the entire 204-acre zoo is an excellent orientation to the zoo.

Special Year-Round Field Trips: These are for school groups, to enhance classroom studies. By reservation only. Please call the Education Department at (708) 485-0263, ext. 365, for a special information packet and reservation forms. (Note: use the 312 area code before 11/11/89.) Special rates apply.

Stroller/Wagon Rentals: Available at both the north and south gates: large strollers for $4.00 with a $1.00 deposit; small strollers for $3.00 with a $2.00 deposit; and wagons for $3.00 with a $2.00 deposit.

Special Exhibits: The following is a list of several special exhibits that land Brookfield in the world-class zoo category: Predatory Ecology; Discovery Center; Australia House—includes kangaroos, hairy-nosed wombats, and Tasmanian devils; Okapis—Brookfield was the first zoo to breed okapis in 1959, and the okapis are worth $30,000; Wolfpack; Hoofed Animals—Brookfield is famed for its hoofed animal collection; and Reptile House.

Souvenirs: The zoo has eight excellent souvenir shops.

Facilities and Access

Hours: Summer hours are 9:30 A.M.–6:00 P.M. Hours change for fall and winter, usually to 10:00 A.M.–5:00 P.M. Call in advance to be sure.

Admission: Adults (ages 12–64), $2.75; children (ages 3–11), $1.00; seniors (age 65 and over), $1.00; and youngsters under 3, free. Free admission on Tuesday. There are additional admission fees to some exhibits.

Transportation and Parking: The zoo is located on First Avenue in Brookfield, Illinois, about 20 minutes west of the Loop, off the Eisenhower Expressway. Exit at First Avenue. Signs will guide you into the zoo. It would be helpful to call (312) 242-2630 and listen to the recorded directions before starting out. Parking fees are as follows: cars, $3.00 (but $3.50 on Tuesdays); buses, $5.00. There is some free parking on the streets, but be prepared to walk.

Restaurants and Food: There are six restaurants within the zoo compound. In the summer, you may wish to picnic on the vast and beautiful grounds.

Restrooms: Well-marked and reasonably clean restrooms abound. Obtain a grounds map when you arrive.

Information: Call (708) 485-2200 or the Chicago telephone number, (312) 242-2630.

CHICAGO PARK DISTRICT

425 East McFetridge Drive
Chicago, Illinois 60605
(312) 294-2200

The single, largest provider of free recreational programs in Chicago for children (and adults) is the Chicago Park District (CPD) with 560 parks and 7,300 acres of land. And, besides the Chicago Public Library, the CPD is probably the most deserving yet least applauded of all city agencies.

The Chicago phone directory under "Chicago Park District" gives you some idea as to the extensive programs available. The CPD Recreation Department alone lists 25 subcategories, including arts and crafts, dance classes, day camps, golf, ice skating, music classes, Special Olympics, tennis, and yoga, to name a few. There are 2 flower conservatories, 7 harbors, 31 natatoriums, and hundreds of parks and playgrounds. There are 235 field houses, 183 gyms, 708 tennis courts, 26 indoor swimming pools (listed as natatoriums), 6 public golf courses, and 2 artificial ice rinks. You are only a phone call away from your nearest park. Literature and brochures can be obtained at local park field houses or at the administration building on McFetridge Drive.

Other facilities of the CPD include Buckingham Fountain, Garfield and Lincoln Park conservatories, Soldier Field, Lincoln Park Zoo, Museum of Science and Industry, Shedd Aquarium, Adler Planetarium, and the Art Institute.

The Park District offers a wide variety of free programs, including the following. Call to find out the exact dates.

Talent Search

Every year in February, at more than 60 parks, prizes of scholarships and music lessons are awarded. Call Dean Goldberg, (312) 294-2320.

Chicago Bluesfest

Usually held in June.

Taste of Chicago

Usually held over the Fourth of July weekend.

Chicago Gospel Festival

Usually occurs at the end of July.

Venetian Nights

Usually scheduled in August.

Chicago Jazz Festival

Usually held at the end of August.

Grant Park Concerts

They begin during the third week in June and continue through August. Grant Park Symphony performs on Wednesday, Friday, Saturday, and Sunday evenings in the Petrillo Music Shell.

Special Neighborhood Concerts

More than 60 per summer.

Flower Shows

Four times a year at both Lincoln and Garfield Park Conservatories: a chrysanthemum show in November, Christmas flower show in December, an azalea show in February and early March, and an Easter show in the spring.

You'll want to see the spectacular year-round flower displays in both conservatories as well as the Main Garden and the Grandmother's Garden at Lincoln Park. The authentic and newly restored Japanese Garden behind the Museum of Science and Industry is spectacular. Located on an island in the Jackson Park lagoons, it shares space with the Paul H. Douglas Nature Sanctuary.

Programs for Physically and Mentally Handicapped Patrons

Call Pat Condon at (312) 294-2330.

Music Lessons and Performances

Contact Dean Goldberg at the McFetridge administration office, (312) 294-2320.

Mom, Pop, & Tots

Special exercise programs for parents and their small children, (312) 294-2317.

While touring the park district's 7,300 acres, take time to note the many monuments and memorials erected and beautifully maintained by the CPD. A complete listing of these, with their locations, sculptors, donors, and dates of unveiling is available at McFetridge Drive. At last count there were 89 monuments.

The Office of Information is extremely helpful, both by phone and in person. Maps, brochures, annual reports, and promotional materials on all facilities and programs are available at McFetridge Drive. The entertainment sections of both papers and *Chicago* magazine carry listings of activities sponsored by the Chicago Park District.

COMMERCIAL TOURS

A few commercial tours are worth taking. We have highlighted them below: some are not usually thought of as tours for children.

Carriage Rides

Horse-drawn carriages are available on the near north side of Chicago, adjacent to the Water Tower Park at the corners of Michigan and Chicago Avenues. During the summer, riding in an open carriage and listening to the sounds of the horse and the traffic is enchanting to children. In the winter, the coziness of the carriage and cuddling under wraps while touring has its own charm.

Bus Tours

The premier bus tour with children is the ride in a vintage, bright red, London-style, double-decker bus. Not only do children enjoy the climb to the upper level, but they love the excitement of having the wind in their faces. Across the street from the Sears Tower at a self-park garage is usually a cluster of the famous red buses. They provide two-hour tours of the city for $5 a person, or $2 for children under 12. Or, you can call the company that runs these tours, the Chicago Motor Coach Company, (312) 989-8919, to charter a bus for a party. The cost is minimal and is well worth it for an overview of the city.

Another inexpensive, adventurous bus ride for a youngster is a ride on a regular CTA bus.

Chicago at Sea: Lake Michigan Voyages

Several companies ply the lakefront and Chicago River with excursions of waterborne delight for even the littlest kids.

Mercury cruises, Chicago's Skyline Cruiseliners, depart from the south side of the Chicago River at Michigan Avenue for one-hour, 90-minute, and two-hour excursions. Call (312) 332-1353 for departure times. Adults $6.00–$9.00, children $3.00–$4.50.

Shoreline Marine Cruises leave from the Grant Park lakefront and usually depart hourly from the Shedd Aquarium until approximately 5:15 P.M. Thereafter, the cruises leave from east of Buckingham Fountain throughout the evening hours until 11:15 P.M. The cruises are approximately 30 minutes in length. Call (312) 673-3399 for details. Adults $4.50, children $2.00.

Wendella Sightseeing Boats can be boarded at the north side of the Chicago River at Michigan Avenue and run for one-hour, 90-minute, and two-hour cruises. Call (312) 337-1446 for schedules. Adults $6.00–$9.00, children $3.00–$4.50.

Chicago From the Lake Cruises: these excursions leave from the North Pier Terminal, 509 E. Illinois Street, and are 90 minutes. Call (312) 527-1977 for information and departure times. This company also schedules an Architecture River Ride cosponsored by the Chicago Architecture Foundation; it includes a light breakfast. Middle-school-aged children and even teenagers (heaven forbid anyone should see them with their [yuch!] parents) would enjoy this jaunt. All participants: $12.00.

Supernatural cruises, boarding at 10:30 P.M. on weekends in the summer from the Mercury dock at Michigan Avenue and Wacker on the south side of the river, have a more limited young audience because of the departure time, but children who are able to stay awake would probably have a lot of fun on this cruise, which runs from 11:00 P.M. to 1:00 A.M. Call Richard Crowe at (312) 735-2530 for information. All participants, $15.00.

More to the taste of the 2- to 11-year-old is the Wacky Pirate Cruise, which departs Friday, Saturday, and Sunday mornings in the summertime from the Mercury Dock at the south side of Michigan Avenue and the Chicago River. Every child must be accompanied by an adult, and everyone gets a kazoo from a pirate in full regalia. Most kids delight in the stories of alleged pirates who cruised our tame shores and love the raucous sing-along. Call (312) 332-1366 for details and reservations Adults $7.00, children $4.50.

A family occasion—Mother's Day, Father's Day, grandparents' birthdays, or such events—can be celebrated on floating restaurants that customarily cater to a more sophisticated crowd. Consider a brunch with your children on the *Spirit of Chicago,* which departs from the south side of Navy Pier at Grand Avenue and the lakefront. Call (312) 321-1241 for information and prices. The *Star of Chicago* runs similar brunch cruises on weekends from the same location. Call (312) 644-5914 for information.

El Rides

As every Chicagoan and many visitors know, the "Loop" of the center city derives its name from the elevated train that rings the city between Van Buren on the south, Wells on the west, Lake on the north, and Wabash on the east. During a non-rush-hour period, the el ride around the Loop and north on the Ravenswood line brings you to a new view of the city that is often overlooked by the scurrying urban dweller. Children can actually look out of the windows and into offices, stores, and even homes on the journey north. It's a safe and inexpensive (though noisy) glance at the underbelly of the city, which most children relish.

AN EVANSTON EXCURSION

Tired of the same summer activities and Chicago crowds? Then hop out of the city for an afternoon and visit neighboring Evanston's beautiful (and less-crowded) lakefront and pleasant museums. Children love having the extra space to play and exploring some new turf. Both the Evanston Historical Society and the Gross Point Lighthouse, which are about two miles apart, are located on (or next to) the lakefront parks.

Evanston Historical Society
225 Greenwood Street
Evanston, Illinois 60201
(708) 475-3410

(Note: use the 312 area code before 11/11/89)

Headquartered in the chateau-esque mansion of the late Charles Dawes, Vice President under Calvin Coolidge, the Evanston Historical Society gives children a glimpse of aristocratic life in the early 20th century. The museum is a bit dark and eerie for children who have fears about old, haunted houses (this would be a great setting for an Agatha Christie mystery), but with a guided tour or some of your own lively explanation of museum artifacts, the trip can be exciting for your children. The museum is neither huge nor geared especially to children, so plan on a short, entertaining visit to the museum.

Before entering, walk around to the lakefront side of the mansion lawn to see the gigantic ship propeller, which was recovered from an early 20th-century ship that sunk 150 feet off the shore of nearby Dempster Street beach. Also, look at the old carriage house that sits behind the mansion. The carriage house was an old, fancy stable and garage used for horses and horse-drawn vehicles. After seeing the outdoor sites, make your way to the entrance on the Greenwood Street side of the mansion.

Once inside, ask if a docent is available for a guided children's tour. If a docent is not available, then you are on your own to investigate the mansion's old sites and objects. The first-floor library is filled with delicate, royalty-style furniture, Oriental rugs, art objects, portraits, and Tiffany lamps. It is definitely not baby- or child-proofed. Be sure to see the white, marble bust of Napoleon that Dawes purchased in 1893 from the French exhibit at the World's Columbian Exposition.

In the Great Hall, also on the first floor, children enjoy seeing the animal trophies of buffalo, caribou, and the African water buck. Most children are relieved to learn that environmentalists today go on photo safaris and not shooting safaris. There is also a first-floor exhibit room focusing on different aspects of Evanston history that older children might find interesting.

After ascending the grand staircase—an adventure in itself—take a left into the doll room where a fantastic collection of early 20th-century dolls is located. An intriguing bathroom filled with all the old bathroom items of previous times is located next to the doll room, and a Tinker Toy exhibit targeted for children is in the planning stages. Also on the second floor are portraits of Native American leaders whose tribes lived in the Evanston area and an old map of the world.

Gross Point Lighthouse
2535 Sheridan Road
Evanston, Illinois 60201
(708) 328-6961

(Note: use the 312 area code before 11/11/89)

Two miles north, along Sheridan Road, is another beautiful lakeshore park in Evanston, this one run by the Lighthouse Park District. It features pleasant open space and a historical lighthouse and nature gallery. The lighthouse is open only on the summer weekends and can only be visited on guided tours (at 2:00 P.M., 3:00 P.M., and 4:00 P.M.). Children are captivated by the stories of the old shipwrecks in the 1800s that motivated the U.S. government to build this lighthouse. The tour is interesting for kids and lets them see the now-obsolete three-wick oil lamp used when the lighthouse was first built, the loud fog siren that is still operable, and the terrific view of Lake Michigan from above.

Don't miss the neat little nature center and nature trail located behind the lighthouse. Staff is available in the nature center to provide discussion and interactive direction with plants, animals, birds, insects, and minerals that are native to the Chicago area and to take children out on the nature trail to explore a piece of suburban jungle.

Evanston Art Gallery

The Evanston Art Gallery also is located right next to the Lighthouse and Nature Center. While its galleries are oriented toward adults, the center does provide excellent summer art classes for children ages 5 to 12. Topics cover painting, drawing, fabric design, photography, ceramics, etc., and the classes are held in superb, professional facilities.

Evanston Lakefront

There are rocks to climb on the shorefront, a lagoon to investigate, a fishing pier to visit, and an ample amount of space, park benches, and playgrounds to make for a very enjoyable outdoor trip. However, beware! If you want to swim, the Evanstonians will hit you up for a $3.00 entry fee to walk onto their beaches, and on holidays and beautiful beach days the Evanston lakefront becomes almost as crowded as Chicago's.

Facilities and Access

Hours: The Evanston Historical Society is open from 1:00 P.M. to 5:00 P.M. daily, except on Wednesdays, Sundays, and holidays, when it is closed. The lighthouse and nature center are open only on weekend afternoons, June through August.

Admission: $2.00 for adults, $1.00 for seniors and children 12 and under, and $4.00 for families at the Historical Society. The lighthouse and nature center are free.

Transportation and Parking: From Chicago, take Lake Shore Drive north to Sheridan Road in Evanston. Sheridan intersects with both Greenwood and Central Streets two and four miles into the city. Ample street and lot parking is available near both the lighthouse and Historical Society. Or, if going by el, get off at Dempster Street in Evanston and walk four blocks east to the Historical Society and lake.

Restaurants and Food: Bring your own and have a picnic. There are also some lakefront vendors selling fast food (hot dogs, ice cream). Otherwise, the college hangouts and fast-food and pizza emporiums of Evanston are a walk south.

Restrooms: There are two restrooms available for public use in the Historical Society.

Information: Call the Historical Society at (708) 475-3410 and the Gross Point Lighthouse at (708) 328-6961. (Note: use the 312 area code before 11/11/89.)

GARFIELD PARK CONSERVATORY

300 North Central Park Boulevard
Chicago, Illinois 60624
(312) 533-1281

Garfield Park Conservatory is one of the finest indoor greenhouses in the country and contains five elaborate and huge greenhouse rooms specializing in various types of fauna: the cactus house, which features numerous prickly dry climate plants; the palm house, which has leafy corners and is excellent for a spontaneous game of hide and seek; the fern house; the warm house; and the economic house.

The conservatory's show halls stage at least four annual events that can be the focus of a visit. In the fall, the chrysanthemum show features all shades of autumn splendor, the Christmas show dazzles the senses with multicolored poinsettias, the late-winter azalea and camellia show portends spring to come, and the Easter show features an imagination full of tulips, lilies, hyacinths, and tender green growing things.

Although the conservatory is located at 300 N. Central Park Boulevard (3600 west) in a less than thriving neighborhood, security is excellent, parking convenient, and the uncrowded conditions warrant a visit.

Facilities and Access

Hours: 9:00 A.M.–5:00 P.M. every day.

Admission: Free.

Transportation and Parking: The Homan Avenue and Kimball Avenue CTA buses stop in front of the conservatory. There's also a free parking lot.

Restrooms: The conservatory's restrooms don't have changing tables.

Information: Call (312) 533-1281.

GRANT PARK

Daley Bicentennial Plaza
337 East Randolph Street
Chicago, Illinois 60601
(312) 294-4790

Grant Park is a huge playground of contrasts for children. The enormous expanse of open field next to the playground defines big and little. Kids can explore public and private space by walking through the crowded areas of the park as well as the quiet areas enclosed by trees. They can see city and nature contrasted by viewing the skyline from near the spectacular wildflower beds.

Your children can examine the joys of land and water firsthand by the availability of paths, bikes, cars, and all the speeds and sounds of Lake Shore Drive, and the lakefront with its boats, sails, and swimmers. This area, between Michigan Avenue (on the west) and the lakeshore (on the east) and between the Field Museum (on the south) and Randolph (on the north), holds more activities for children and their parents than just the traditional festivals, which make Grant Park and summertime in Chicago synonymous.

Sights

The nightly light shows at Buckingham Fountain are great fun for people of any age. They run from May 1 through October 1, beginning at 9:00 P.M. and ending at 11:00 P.M.

From early spring until early fall, don't miss Chicago's largest wildflower display, in the Daley Bicentennial Plaza (bordered by Monroe, Randolph, Columbus, and the lakeshore). The beauty of the flowers is paralleled only by the skyline view. For the nature-loving and bird-watching child, note that this area of the park attracts numerous land birds, just as the lakefront draws a variety of gulls to its shoreline.

Sports

Skating is offered year-round at the roller rink, which converts to an ice rink in the winter. The rates are $.75 to rent children's skates and $.75 for rink time; the comparable rates for adults are $1.25. A beautiful bike path runs parallel to the lake—another excellent way to view the skyline and experience the cooling lake breezes. You and your children may also want to try the free tennis courts open by reservation on a bring-your-own-equipment basis. The Daley Bicentennial Plaza also includes a terrific playground for kids who have a zest for free play instead of organized sports.

Toddler Programs

The Daley Bicentennial Plaza offers toddler programs in 10-week sessions from fall to spring. The Moms and Tots program is organized for kids ages one to three accompanied by a parent or substitute. The emphasis is to get kids out with other children in the serene company of their parents or caretaker. Your children can meet other kids and parents while learning to participate in group activities. The kids practice motor control and dexterity in a comfortable group with their peers. The cost for 10 weeks is either $10 for once-a-week visits or $20 for two sessions per week. The Kiddie College program is for kids ages three to five and focuses on learning skills: the ABC's, motor control, numbers, and working within a group. These programs are scheduled during fall and spring. For more information, call the Daley Bicentennial Plaza at (312) 294-4790.

Music/Festivals

In addition to the famous Bluesfest in early summer and the Taste of Chicago in mid-summer, Grant Park offers a wonderful opportunity for your child to hear classical music—for free. The Petrillo Music Shell concerts begin in late June and continue through the end of August on Wednesday, Friday, Saturday, and Sunday nights. After a picnic in the Daley Bicentennial Plaza lawn or seating areas, the Petrillo Music Shell is a comfortable atmosphere in which to share and enjoy your child's first classical concert under the stars.

Facilities and Access

Hours: Park grounds are open from daylight to 11:00 P.M. The Daley Bicentennial Plaza itself is open from 8:00 A.M. to 7:00 P.M. daily.

Admission: Free.

Transportation and Parking: Parking is nearly always available in the Grant Park underground garages, which are operated by the Park District. Metered parking is less frequently available along Columbus Drive. Here are some telephone numbers for garages: Grant Park North Garage, (312) 294-4598; Grant Park South Garage,

(312) 294-4593; Monroe Street, (312) 294-4740.

Restaurants and Food: Concession stands and beverage machines in the buildings; vendors during warmer months.

Restrooms: Locker rooms for men and women in the Daley Bicentennial Plaza, 337 E. Randolph, provide ample room for baby changing. The public restrooms in Grant Park are questionable at best. Go before you reach the park.

Information: For information about Bluesfest, Taste of Chicago, and Gospelfest, call the Special Events line, (312) 744-3315. For information on free classical music concerts, call the Petrillo Music Shell, (312) 294-2420. For other information call the Daley Bicentennial Plaza, (312) 294-4790.

JACKSON PARK

6401 South Stony Island Avenue
Chicago, Illinois 60637
(312) 643-6363

Summer excursions to Lincoln Park to see the zoo or Grant Park to attend a festival are annual trips for most northside Chicagoans. But you may decide that *you* would like to visit a new sight. So for your sake, and not just the children's, venture down to the not-so-hidden treasure of the South Side, Jackson Park. (Like most Chicago parks, it's not advisable to visit after dark.)

During the day, however, Jackson Park is safe and beautiful and offers a variety of features. Among the many recreational facilities in the park are children's playgrounds, baseball and softball fields, basketball and tennis courts, an 18-hole golf course, a driving range, paddleboat rentals, and three beaches. Grant and Lincoln Parks may snatch most of the publicity, but Jackson Park has as much to offer as both those parks—minus the crowds and parking hassles.

Columbian Exposition

The spectacular World's Columbian Exposition of 1893 was staged in Jackson Park. Many of the buildings and gardens are remnants of that great fair. The former Palace of Fine Arts, now known as the Museum of Science and Industry, is the Exposition's most famous legacy. Three grand monuments are also a reminder of earlier times. The famous Fountain of Time, the Gotthold E. Lessing monument, and the sculpture of Carl Von Linne will awe children with their size and grandeur.

Geography of Jackson Park

The 543-acre park lies along the Lake Michigan shoreline and is bordered on the north by the Museum of Science and Industry and on the west by Stony Island Avenue. In addition to Lake Michigan, the park has picturesque lagoons and a pleasant yacht harbor. South of the museum lies Wooden Island, a nature sanctuary located in the park's largest lagoon. Show your children the beautiful flora and fauna and see if you can locate some of Illinois's more exotic birds, like the sandhill crane, the evening grospeak, and the green heron. If you are not a nature expert, try to hook up with Doug Anderson on one of his early-morning bird-watching walks that start in the spring and run through fall (call [312] 493-7058, weekdays between 5:00 P.M. and 6:00 P.M. only). Don't miss the Japanese gardens at the north end of the island, which were a wonderful gift from the Japanese government for the 1893 exposition.

In the west-central section of the park, an elaborate formal garden annually displays more than 180 varieties of flowering plants. Its floral displays, maintained from April through fall, include masses of tulips, hyacinths, narcissus, and chrysanthemums.

Jackson Park Field House

Located on the 6400 south block of Stony Island Avenue, the Jackson Park Field House was built in the 1950s on the same plan as the Russell Square and Avalon Park field houses. It has basketball courts (with tough competition), an 18-hole golf course built in 1900 ($5.50–$7.50, call [312] 493-1455), and occasional programs for children. In the winter, the golf course area maintains a skating rink and is an excellent place for cross-country skiing. There is also a driving range at Hayes and Lake Shore Drive.

Washington Park Field House

The Washington Park Field House, located approximately one mile west of Jackson Park, is actually more suited to the needs of children than the Jackson Park Field House. It is a much more modern building with a full schedule of activities for children and adults. There are tennis, racquetball, and handball courts; football and softball leagues; woodworking, ceramics, and photography shops; music and drama classes; and a lot more. For a full rundown and a schedule, call (312) 684-6530 or write the Washington Park Field House, 5531 South King Drive, Chicago, Illinois 60637.

Jackson Park (and Washington Park) is an exciting new area to explore for adults and children alike. Go there to picnic and walk through the Japanese garden, check out the field-house games, and show your children how to hit a few balls at the driving range. And if you are a cross-country-skiing buff, come in the winter and explore one of Chicago's more fascinating parks.

Facilities and Access

Hours: Jackson Park is open from daylight to 11:00 P.M. The Field House is open from 8:00 A.M. to 10:00 P.M. The first classes begin at 1:30 P.M.

Admission: Free.

Transportation and Parking: By car coming south, stay on Lake Shore Drive past the exit for the Museum of Science and Industry. Turn right at the next stoplight. Parking is available in several lots surrounding the park. The Stony Island and 63rd Street buses reach the park, but they don't go through it. To reach the golf course, get off at 63rd Street and walk east a few blocks.

Restaurants and Food: During the summer months, concession stands supply basic snacks, such as hot dogs. Consider going to the Museum of Science and Industry.

Restrooms: Each field house has adequate facilities for babies, children, and adults.

Information: Call the Jackson Park Field House, (312) 643-6363.

LINCOLN PARK AND THE LAKEFRONT

Field House and Cultural Arts Center
2054 North Lincoln Park West
Chicago, Illinois 60614
(312) 294-4750

Lincoln Park is bounded by Cannon Drive on the west, Diversey on the north, LaSalle Street on the south, and the lake on the east. For you Chicago experts, here's a little quiz. Which of these can be found in Lincoln Park?

900 acres of man-made landfill
Free children's tennis lessons
Zebras, otters, and flamingos (see Lincoln Park Zoo)
A bicycle and roller-skate rental outlet
Goat milking and sheep petting (see Lincoln Park Zoo)
A miniature golf course
Sea lions and penguins (see Lincoln Park Zoo)
Children's kite-flying and car-racing contests
A half-dozen playgrounds equipped with baby swings
A boat club that boasts a junior sailing program
A 40-foot totem pole and 23 other outdoor sculptures

Yes, the answer is all of the above.

So you can see, Lincoln Park is not just a jogger's haven. With more than 1,200 acres of grassland, parks, sculpture, trees, lagoons, harbors, statues, benches, and bike trails, the park indeed is a wonderful place to jog or simply stroll with a child, but it offers much, much more.

Dig a little beyond the surface and you'll find a myriad of activities designed especially for youngsters. For aspiring golfers, there's a nine-hole course, a driving range, and a miniature golf course. There are dozens of tennis courts and free lessons for children 17 and under. For those children who can't get enough of the water, the Chicago Yacht Club runs a top-flight (but expensive) Junior Sailing Program. And then, of course, there's always the Lincoln Park Zoo (see the Lincoln Park Zoo section). Below is a more detailed description of some of the major attractions within the park.

Open Park Space

The vast, undulating stretch of land that we now call Lincoln Park was once just a small cemetery, called the City Cemetery. In fact, the park only came into existence because Chicago's health experts in the 1860s believed that decomposing bodies in the cemetery gave off deadly, noxious gases. In order to solve this health problem, the city fenced off the cemetery's unused northern half as a park (but not before 1,000 Confederate prisoners from Camp Douglas on the south side of Chicago had been buried there). Due to the popularity of this park, in later years the city of Chicago dug up the bodies, resod the soil, and began to extend the park northward by landfill. The remains of the soldiers were transferred to Calvary, Rosehill, Grace, and Oakwood (where a monument stands to their memory) cemeteries. Today, fully three-fourths of Lincoln Park is man-made (from Diversey Avenue to Ardmore).

Because of far-sighted planning, Chicago children and parents now have an immense playground along the shore of Lake Michigan. Feeling cooped up and over-cartooned, with fretful children at hand? Take a stroll in Lincoln Park. Stop and watch the Hispanic, or the Iranian, or the Haitian soccer teams get in condition. Walk over to the Totem Pole and guess how tall it is (answer: 40 feet). Head down to the lagoon and watch the ducks skim the water and the remote-control boats whiz back and forth. Or just sit on the edge of the shoreline and watch the boats gliding into Belmont or Montrose harbor, the waves lapping up against the rocks, and the breeze blowing. If you have a little more time, rent a paddleboat, or bring along the children's bikes or roller skates (you can rent adult-sized bikes or roller skates in the park, see below). Or if you have a little less energy, visit one of the three cafes in the park.

Lincoln Park Bicycle and Roller-Skate Rental

If you exit Lake Shore Drive at Fullerton, you'll see this outlet on the left at the first stoplight (2400 North Cannon Drive, [312] 294-4682). It's open weekends in April and May, daily from Memorial Day to Labor Day, and weekends in September and October—weather permitting. Hours are 10:00 A.M.–6:30 P.M. (no rentals after 4:30 P.M.).

The bike-rental fees are $4 for the first hour and $3 for each additional hour (one-hour minimum). You need to provide a deposit of a driver's license and a major credit card or $50 cash. No children's bikes are available for rent.

The skate-rental fees are the same, but the required deposit is less: driver's license and major credit card or $10 cash. The skates come in men's sizes 4–12.

Paddleboat Rental

During warm-weather months the zoo operates a paddleboat concession for the South Pond behind the Cafe Brauer (about 2000 north on Stockton, just north of the Farm-in-the-Zoo). The large paddleboats hold five people, and the small ones hold three. Children must be tall enough to reach the pedals and must wear life preservers, which are provided. Call the concession directly for hours and costs: (312) 871-3999.

The park district operates a paddleboat concession for the North Pond (Stockton Drive just north of Fullerton Avenue).

Outdoor Sculpture

There are 24 pieces of outdoor sculpture in Lincoln Park and north along the lakeshore. Following are some of the kids' favorites.

The totem pole at 3600 north is also called *Kwa-Ma-Rolas.* The founder of Kraft, Inc., James L. Kraft, donated this sculpture to the schoolchildren of the city in 1929. To this day, the sculpture's vivid colors and exotic look attract the gazes of children who pass by on foot or by car on Lake Shore Drive.

Storks at Play, next to the Conservatory, depicts the sheer joy of playing in the water. Consisting of boys and swans playing in the water, this sculpture is so moving that the city had to place a fence around it to deter children from joining in the fun.

The Alarm (3000 north), the oldest outdoor monument on park district land, honors the first Native American settlers in the area, the Ottawa Nation. Children enjoy looking up at the Native American family.

Lincoln Park Field House and Cultural Arts Center

The Cultural Arts Center, 2054 North Lincoln Park West, (312) 294-4750, offers a variety of inexpensive classes and contests for youngsters. In the fall and spring, the center provides aerobics, woodcraft, jazz, tap-dancing, and lapidary (jewelry-making) courses. The lapidary class is particularly unusual, as children use a considerable amount of advanced equipment—stone cutters, shavers, etc. The instructor of the lapidary class, John Ferguson, has been at the center for more than 16 years. The cost of most of these courses is just a $10 registration fee.

The center also supervises several annual contests. In May, a children's kite-flying contest is held in the park. Center staff help children build their own kites in the preceding months and then judge the contest. During the summer there is also a car-racing contest for kids who have built their own cars. The center's woodcraft instructor assists the youngsters in building their dream machines.

Besides the courses and contests, the center operates an inexpensive summer camp for children ages 8–12. The camp runs five days a week for approximately two months. Activities include aerobics, sports, crafts, and swimming. Current price is $40. Call Tony Mendocino at (312) 294-4750 for details.

Lincoln Park Conservatory

The conservatory is located at 2400 North Stockton Drive, (312) 294-4770. If your child believes there are only two kinds of plants—grass and the kind you spray weed-killer on—then you owe him or her a visit to the Lincoln Park Conservatory. Offering a wonderfully relaxing environment that features classical music and numerous penny-pitching ponds, the conservatory also boasts one of the finest horticultural collections in the world. While the conservatory is only for display and there are minimal tours and activities, many parents enjoy simply strolling through the four large, glassed-in buildings—the Palm House, the Fernery, the Cactus House, and the Show House. In the Show House, the park stages seasonal shows to anticipate the season. Traditionally, there is a spring azalea show, a fall chrysanthemum show, and a winter poinsettia show with other special features.

Usually the hours are 9:00 A.M. to 5:00 P.M., but sometimes they are 10:00 A.M. to 6:00 P.M. Call to be certain. Some children enjoy seeing plants they recognize from home, others prefer to run over the bridges and toss pennies in the ponds. Although the conservatory will not provide a day's worth of activity, it is a soothing respite from the concrete, steel, and plastic of the city. It's also next door to the Lincoln Park Zoo, so pop on over if you want to get away from the noise or the smell of the animals. Admission is free and restrooms are near the entrance.

Golf

For young golfers, there is a nine-hole course at Waveland Avenue. Juniors (ages 16 and under) can play for $2.50 if they have a junior card (the adult greens fee is $5.00 during the week, $6.00 on weekends, plus $1.00 more for nonresidents). To obtain a junior card, simply bring $1.00, a small photo, and some proof of age to the starter's shack. The starter will make your card while you wait. The course is average, but it is nicely situated along Lake Michigan with a good view of the city. Call (312) 294-2274 for more information.

There is also a driving range just north of Diversey Avenue. The range is open to all ages, every day from 8:00 A.M. to 10:00 P.M. (April 1–November 10). Three dollars will buy you a large bucket of balls. For those youngsters seeking lessons, a handful of golf professionals list their qualifications and phone numbers on the bulletin board inside the driving range office. These lessons take place at the range and cost around $20/half hour. Call the Diversey Range at (312) 281-5722 for more information.

Adjoining the Diversey Range is a spartan miniature golf course without gewgaws or frippery—just putting. It's an excellent place to take a group of 10-year-olds seeking mastery of this preadolescent passion. The fees are $2.50 for adults and $2.00 for children ages 12 and under. Hours: 7:00 A.M.–9:30 P.M.

Tennis

Waveland Courts: Waveland has free, once-a-week tennis lessons during the summertime for children ages

17 and under. Register in person at the trailer next to the courts. On Tuesdays there is a Junior Development Program for advanced youth, and on Fridays at Waveland there are youth matches against other park teams. Twenty public courts, $2 daily permits for adults. Call (312) 281-8987 for information—try calling in the afternoon and let the phone ring a long time. If no one answers, try (312) 294-4792.

Diversey Courts: Next to the golf driving range just north of Diversey Avenue, there are four public tennis courts. If you want to play with your children, you must register in person at the public courts (cost: $7/hour).

For private children's lessons, contact the Mid-Town Tennis Club, 2020 West Fullerton Avenue, Chicago, Illinois 60647, (312) 235-2300.

Wilson Courts: At the Wilson Courts there are five free public courts. A rack-up reservation system is used.

Sailing

The Chicago Yacht Club sponsors a junior sailing program at the Monroe Harbor. The curriculum of this program includes: water safety skills, understanding equipment, and skills in handling small boats (both sail and power). The program runs Monday–Friday, 8:30 A.M. – 4:30 P.M., and the cost is $725 for eight weeks and $375 for four weeks (members receive a discount). Call (312) 861-7777 for more information.

Skeet-Shooting

The Lincoln Park Skeet and Trap Range offers skeet-shooting lessons for youngsters who can read and write and are old enough and strong enough to handle a shotgun safely. These semi-private classes focus on the use of a shotgun, and are offered only in the spring and summer. Each class is comprised of four two-hour sessions: the first session uses a movie and classroom visual aids to stress safety; the last three sessions are supervised practice on the outdoor range. The price of this course is $120 and includes targets, shells, plastic safety glasses, hearing protection, and a lengthy beginner's pamphlet. Club guns are available for adult classes, but children must bring their own guns, appropriate to their body size. For more information, call (312) 549-6490, and ask for Ed Bishop, the manager, or Judy.

Skating

During the winter months, the Chicago Park District floods the baseball fields to the east of Brett's Cafe. The Park District creates two rinks, each about the size of a football field, and designates one for hockey and the other for figure skating. These rinks are usually open from 6:00 A.M. to 11:00 P.M.

Theater

Theater-on-the-Lake, at 2400 north, does not offer special children's productions. However, the theater occasionally puts on family-type plays, like *South Pacific.* If one of these plays is running, the theater is a very relaxing and inexpensive place to bring children. There are no reserved seats, so arrive early to place your mark and then stroll the evening lakefront until showtime. Call (312) 348-7075 for the current schedule.

Friends of Lincoln Park

Friends of Lincoln Park is located at 990 West Fullerton Avenue, Chicago, Illinois 60614, (312) 525-1853. It's an excellent source of information about children's activities in the park. They are in the process of designing a variety of new programs (i.e. cross-country skiing and a children's nature/historical tour of the park) and are happy to speak to the public.

Facilities and Access for Lincoln Park grounds

(See separate listings for golf course, skeet-shooting range, etc.)

Hours: Park grounds open at daylight and close at 11 P.M.

Admission: Free.

Transportation and Parking: Free and metered parking throughout the park.

Restaurants and Food: Brett's Cafe is located at the Waveland Avenue golf course—open from March through October, weather permitting, 8:30 A.M.–4:00 P.M. weekdays and 7:30 A.M.–4:00 P.M. weekends, (312) 528-6888. Brett's is a cozy, family-oriented cafe. For kids who want the all-American favorites, Brett's offers hot dogs, french fries, and grilled cheeses. But for the more adventurous youngsters (or adults accompanying them), Brett's has a number of unusual, healthy treats: Swiss, avocado, and tomato sandwiches; half-buckwheat and half–whole-wheat pancakes; bran and granola cereal; homemade chili; and homemade cake. The staff is very friendly, and the menu, written in colored chalk on a blackboard, catches kids' eyes.

Cafe Brauer, located on Stockton Drive about 1900 north, is the cafeteria for the Lincoln Park Zoo. Loud, noisy, and a bit grubby, the cafe caters to children and families, cafeteria-style. When the leaves are falling from the trees in the autumn, take your children to the second floor of the cafe to enjoy the beautiful view of Chicago.

The cafe is open daily 10:00 A.M. to 5:00 P.M. during the warm months, and Monday–Friday 10:00 A.M. to 3:00 P.M. and Saturday and Sunday 10:00 A.M. to 4:00 P.M. during the rest of the year. Call (312) 280-2724 or (312) 935-6700.

Park Place Cafe is hidden away in Lincoln Park at the north end of North Pond, south of Diversey between Stockton and Cannon Drives across from Columbus Hospital. Accessible only on foot, the cafe is open eclectic hours and has no telephone. Stop by and see. Park along Stockton Drive, Cannon Drive, or Lakeview. Their chili has numerous fans in the surrounding community who even carry it out for football afternoons.

Restrooms: Public restrooms are located throughout Lincoln Park's grounds, but they are not recommended. Your best bets are the facilities at the Lincoln Park Zoo—in the Children's Zoo, Large Mammal House, and Hoof Stock House.

Information: Call the Lincoln Park Field House and Cultural Arts Center, (312) 294-4750.

LINCOLN PARK ZOO

2200 North Cannon Drive
Chicago, Illinois 60614
(312) 294-4660

Zoos appeal to a child's wild imagination, and Lincoln Park Zoo is no exception. Home to more than 2,000 animals, birds, and reptiles, this is a world-class zoo with all the penguins, gorillas, gazelles, hippos, and polar bears you or your youngster could ever want to see. But Lincoln Park Zoo also houses numerous exhibits especially for children. Following is a rundown of these special exhibits.

Children's Zoo

This is a special zoo within a zoo, presenting animals on a child-sized scale. In the Animal Gardens, for example, hedgehogs and river otters jump and wiggle about in miniature habitats that are easily accessible to kids. The signs are even child-oriented, standing two feet off the ground and posing such common-sensical questions as, "Why can otters swim better than you can?" and "These animals like to eat fish best of all. Do you?" All in all, this is a wonderful opportunity for children to get close to some of the wildlife and practically stick their noses into the cages and habitats. At the Hands-on Zoo, petting the

African pygmy goats is actually encouraged.

Another exciting place within the Children's Zoo is the Kids' Corner: A Discovery Place. Designed as an educational environment for children under 12, this room is chockful of hands-on displays: a magnified view of snake skin, a steel brush used on elephants, zookeepers' tongs used to reach into unfriendly cages and grab lunch bowls, and a tunnel full of buttons and flashing lights that examines the daytime and nighttime lives of wild animals and insects. An expert from the museum is always on hand, but call (312) 294-4649 in advance to make sure that a school group has not reserved the area. The Kids' Corner is usually available to families on the weekends and between 2:45 P.M. and 4:30 P.M. on weekdays.

Wildlife Workshops

Ever hear of Blossom the Opossum and Arnie the Armadillo? These are two of the stars featured in "Animal Hop, Skip, and Jump," a special program designed to introduce toddlers to the different ways that animals move. Essentially, these workshops attempt to explore in depth one specific aspect of life at the zoo. Past seminars have included: "Wet and Wild," an examination of aquatic creatures; "Now You See Me, Now You Don't," a study of some of the best camouflage artists at the zoo; and "Barnyard Babies: How Farm Animals Care for Their Young," that includes a hands-on investigation of piglets, chicks, and calves. These workshops usually last one to two hours and range in cost from $3 to $11.

During July and August, the Children's Zoo also holds several special, one-week summer camps for children ages 8–10. In the past, these camps have revolved around adventure-oriented activities: nature hunts, storytelling hours, and visits behind the scenes to get a firsthand view of a zookeeper's life. For information on these camps or any of the workshops, call (312) 294-4649 to inquire about the Education Program for Students.

Everyday Special Events

"There's something happening every minute at Lincoln Park," claims the Lincoln Park Zoo brochure. And it's true. Every 15 minutes at the zoo, there is some type of special event that appeals to children: goat milking, elephant workouts, ape feedings, polar bear films, and "Animal-of-the-Hour" and "Meet-the-Animals" presentations. Be sure to get a listing of the day's special events from the information booth near the main entrance.

Farm-in-the-Zoo

For those kids (or parents) who think vegetables grow on the shelves of the supermarket, Farm-in-the-Zoo is a real eye-opener. This unique section of the zoo allows youngsters to literally get a feel for what life is like in the country. At the Main Barn Learning Center, children hold baby chicks, pet calves and lambs, and try their hand at milking a goat ("Bonnie" and "Brian" are the names of the twin goat-kids presently residing at the farm). Children also transform themselves into farmhands by placing their heads into a cutout titled "Picture Yourself on the Farm Team" (a display guaranteed to elicit a "Look, Mommy! Look, Daddy!" from your child). The Learning Center also oversees farm-related arts and crafts activities. Although the farm is very busy, there's no need to worry that you'll miss an event because the Farm-in-the-Zoo staff announces special activities once an hour over the loudspeaker.

After leaving the Main Barn Learning Center, be sure to visit the other miniature red barns. In the Poultry Barn children line up two or three deep around an incubator, where chicken eggs rock and baby chicks pip out of their shells, with their eyes shut and their feathers matted down. At the Horse Barn, kids see real live Clydesdales, mules, and pinto horses strutting about the corral or being groomed in their stalls. Inside the barn, kids answer a special horse quiz (i.e., how tall was the first horse? 14

inches high). Another highlight of the farm is the Farming is Feeding garden.

Facilities and Access

Hours: 9:00 A.M.–5:00 P.M. every day.

Admission: Free.

Transportation and Parking: The Lincoln Park Zoo is located between Cannon and Stockton Drives just south of Fullerton. Exit Lake Shore Drive at Fullerton and turn left at the first light onto Cannon Drive, which is a one-way, metered parking street. Or park for free along Stockton Drive if you can find a spot. You can reach the zoo on CTA buses 151 Sheridan and 156 LaSalle.

Restaurants and Food: Cafe Brauer (see Lincoln Park), Landmark Cafe next to the Seal Pond, and numerous vendors.

Restrooms: Restrooms are located in the Children's Zoo, Large Mammal House, Hoof Stock House, and nearby in the Lincoln Park Conservatory.

Information: Call (312) 294-4660 for a recorded message and directions. Call (312) 294-4662 for more information. Call (312) 294-4649, the Education Department, to arrange group tours, classes, and traveling zoo appearances.

NATURE CENTERS

For children who love nature centers and the myriad of facts they can gather up and clutch, there are numerous nature centers in the suburban areas that bear investigation.

Within the city, North Park Village Nature Center at 5801 N. Pulaski, (312) 583-3714, provides some of that country flavor. In the early spring—maple sugar season—the center taps trees, boils down its own syrup, and hosts a delicious pancake breakfast. Also, on the weekend before Halloween, the park and village put together a night walk in the woods to visit Halloween visitors portrayed by volunteers dressed in scary costumes, carrying running buzzing saws, and stirring huge caldrons. While the event may be a little scary for small children, there are so many people there joking with the spooky creatures that real fear evaporates and instead, your children can experience the thrill of squealing in the night forest, clutching your familiar hand.

6
CHICAGO'S NEIGHBORHOODS

ANDERSONVILLE—CHICAGO'S LAKE WOBEGONE

When the King and Queen of Sweden came to Chicago in the Summer of 1988, the city turned Clark Street, from Foster to Balmoral, into the festival route, and the Andersonville neighborhood was decked out in traditional Swedish blue and yellow to enjoy a day-long parade and party.

Even when kings and queens don't come to Chicago, your little princes and princesses may enjoy a great, day-long festival in July held in the old Swedish community. Try to visit the neighborhood during this festival time, but if you can't, do stop sometime in this pleasant area. The local residents seem untainted by big-city life and a courteous "hello" or "good morning" seems customary for this part of town.

Restaurants and Grocery Stores

There are two excellent Scandinavian restaurants in the area to choose from: Ann Sather's, 5207 North Clark, (312) 271-6677; or Svea, 5236 North Clark, (312) 334-9619. For breakfast, try Svea. This little establishment knows how to prepare a perfect morning meal. The Swedish pancakes are simply the best in Chicago. If the Swedish pancakes are too sweet for your taste, try some eggs and compliment them with some limpa bread and Swedish potato sausage. The seating area is small and intimate.

If you are visiting the area in time for lunch or dinner, try Ann Sather's. They have recently moved into the neighborhood and their menu offers something for everyone. Give the children a chance to taste a variety of Swedish delights by ordering the Swedish Sampler. The sampler comes with duck breast, lingonberry glaze, meatballs, potato dumplings, sauerkraut, and brown beans. The desserts are excellent: take your pick between pies, pudding, or Ann Sather's trademark cinnamon rolls.

If you enjoy Swedish cuisine and would like to try your hand at making dishes at home, stop in at Wikstrom's Gourmet Foods, 5247 North Clark, (312) 878-0601 (north of Ann Sather's). Four Scandinavian flags and the U.S. flag fly in front of the shop. Your children will see some old-time Swedish residents who enjoy hanging out and chatting in the seating area. Try the complimentary and aromatic coffee and peruse the shelves for Swedish pancake mix, bread dumpling mix, or maybe some Scandinavian cheese. If Wikstrom's does not have something you need, go across the street to Erickson's Delicatessen and Fish Market, 5250 North Clark, (312) 561-5634. Between the two stores you will find whatever you need for an authentic Swedish meal at home.

Bakeries

There are two Swedish bakeries in Andersonville. On the Foster end of Clark Street is Nelson's Scandinavian Bakery, 5222 North Clark, (312) 561-5494. They have a wide selection of Swedish sweet breads along with some almond pretzels and pecan frisbees. Near the corner of Balmoral and Clark Streets is the Swedish Bakery, 5348 North Clark, (312) 561-8919, which has a slightly larger and more delicious selection with some specialties, such as the Danish almond coffee cake and the princess tart. Your children may also like the red-rimmed butter cookies. The Swedish Bakery is mobbed just before a holiday. Call ahead to order and get there early! Both bakeries are definitely worth a quick visit.

Special Stops

Time permitting, the Swedish Museum (see Ethnic Museums section), located next to Ann Sather's, offers beautiful displays that demonstrate to you and your children the history and pride of the Swedes. If your kids are book lovers, then visit the Fiery Clockface Bookstore at 5311 North Clark, (312) 728-4227. Their special children's section is equipped with kid-sized seats. Both stops are sure to entertain the curious adventurer.

Facilities and Access

Transportation and Parking: The Foster Street el is a long walk from Andersonville, so it is best to visit by car or by one of the Clark Street buses. By car, take Lake Shore Drive to Foster and drive a mile westward to Clark Street. There is metered parking on Clark Street, but open spaces are scarce. Try parking on nearby Ashland Avenue or on one of the lovely adjacent side streets.

Information: For information about street fairs, holiday celebrations, and other events, contact the Andersonville Chamber of Commerce at (312) 728-2995.

ARGYLE STREET—LITTLE SAIGON

Taking your children for an afternoon through Argyle Street's Vietnamese and Chinese neighborhood is a wonderful way to taste the culture of another world. This area spans only four city blocks, but exploring that small space can take a day and can almost give your children the experience of foreign travel.

Children (and adults) are excited to discover different foods, stores, and people right here in the middle of the sometimes-bland Midwest. The community is extremely friendly and polite to outsiders—especially to little ones. The opportunity we have to share in their culture is a real delight. A weekend afternoon is prime time for a family visit because many neighborhood families are out and about doing their weekly shopping. Your children can mix with the Vietnamese children and turn the local supermarket into their own private playground.

Restaurants and Bakeries

Some of the great pleasures of ethnic neighborhoods are their restaurants. This area is no exception. Only Chinatown has a larger selection of Oriental food than Argyle Street. If your children are the "I'll only eat hamburgers" type, try plying them with scrumptious Oriental cookies. If they are more daring, you can tempt them with a whole new world of seafood, chicken, pork, and beef dishes prepared in new ways.

The Hue Restaurant, a half block west of the el (1138 West Argyle, [312] 275-4044), is a good kids' stop. The atmosphere is homey and the service people are pleasant, willing to answer all questions, and good to children. The restaurant is not purely Vietnamese, however. It refers to its style as "Vietnamese cooking with a French influence" (the influence seems minor). Be sure to taste their specialty, the asparagus crab soup, and try some of the outstanding seafood, including pineapple shrimp, red snapper, and steamed pike. Some of their more exotic dishes like mu shu beef, orange chicken, and coconut pork are exciting to eat because the Oriental food is assembled at the table, with strange-smelling sauces—all good fun.

Across the street is the more authentic Lac Vien Restaurant, 1129 West Argyle, (312) 275-1112, which has an extensive menu at reasonable prices ($4–$8). Great seafood and vegetable selections lead the list, which also includes some selections for eating daredevils, such as shallow fried eel, bamboo duck soup, and spicy catfish soup.

Two blocks east of the el are two other excellent restaurants, Nha Trang's, 1007 West Argyle, (312) 989-0712, and Nhat Phuong's. Try the moo goo gai pan and the kung po shrimp at Nhat Phuong's and chat with the friendly owner there. The lemon iced tea at Nha Trang's is great on a sweltering summer day. All dishes at Nha Trang's are excellently prepared and served. Your children might squeal at the thought of eating the fried frog legs with lemon grass.

A block west from Nha Trang's is the New Hong Kong Bakery, the place for dessert. Be daring. From the coconut balls to the spice cookies, everything sampled is a little bit surprising, but always has a splendid taste. Most children will love the shape, taste, and color of the butterfly and almond cookies—and the cream horn will knock their socks off with sweetness. The counter people seem partial to children and are ready to make special suggestions if you are having trouble deciding what goodies to taste.

Viet Hoa Market

You may forget you are in Chicago when you step into the Viet Hoa Market, 1051 West Argyle, (312) 334-1028. It appears to be transplanted straight from Saigon. As you enter, inhale the intense, mixed aroma of soy, spices, and whatever else occupies the market space that day. Thrown into temporary confusion by the apparent chaos in the aisles and the line to the register, you might have the urge to flee. Remember, though, you don't have to shop. It's fun for your children just to look and marvel at a market that bears little resemblance to the sterile and organized clone supermarkets that are scattered throughout our city.

First, view the fish display and see the live crabs, the squid, and the huge frozen eels. Proceed to look at the vastly different produce section and see if you or your children can identify some—or any—of the Oriental and intriguing vegetables. Then listen to the intercom that broadcasts its messages in Vietnamese, not English. And while your child is captivated by the sounds and sights, slip over to the tea section and select some excellent jasmine tea (or any of the other flavors that interest you). The first aisle is worth a stop, where inexpensive but beautiful kitchen utensils are sold—the likes of which may someday be emulated in more fashionable shops. If you are feeling guilty at the end of the visit for leaving the market empty-handed, think about buying the sweet peanut wafers, which are a great and arguably healthy snack.

Medicine Store

After leaving the Viet Hoa Market, walk a block east to the Lins Trading Company, 1011 West Argyle, (312) 271-9714, the pharmacy for the Vietnamese. Go to the shop owner and point to your stomach, your head, or your feet and describe an imaginary (or real) discomfort you are feeling. He will comb the store looking for some ginseng, a fish fin, or some herbal remedy or concoction that may do the trick on your ailment. All the mysterious medicines displayed may be more effective than we or our children imagine—and it's fun to look.

Viet Nam Museum

The Vietnamese community in the Argyle Street neighborhood has grown in just the last 10 to 15 years. The area has its own museum (located one block west of the el at 5002 North Broadway, Chicago, Illinois 60640, [312] 728-6111) and some excellent neighborhood development organizations. The Vietnam Museum—opened recently by a Vietnam veteran—is a collection mostly of Vietnam War memorabilia: pictures, maps, uniforms, military gear, and peculiar paraphernalia that will capture the attention of older children and teach them a bit of history. The museum is a good final stop on the Argyle Street tour. The museum is open Tuesday–Friday, 1:00 P.M.–5:00 P.M.; Saturday and Sunday, 11:00 A.M.–5:00 P.M. Admission is free.

New Year's Celebrations

Chinese New Year celebrations are a welcome break from the dreariness of Chicago winter. Since the Chinese calendar is based on a lunar cycle, the New Year falls on a different day every year, usually between January 20 and February 20. Since the Asian cultures are family- and child-centered, including all children in the holiday celebration is easy. It's well worth a trip into the Asian communities of Chinatown or Argyle Street to participate in the festivities.

Each Chinese year in a 12-year cycle is named for an animal—1989 is the Year of the Snake. A traditional celebration is a parade and lion dance in which a 20-foot-plus, radiant, lion-headed dragon wends its way down the main streets of the Asian districts, stopping and bowing at shops and homes to visit and bring good luck for the coming year.

Your school-age children may have read that Chinese parents and relatives give gifts of money in red envelopes to children and ply them with special rice cakes and holiday sweets. Most children know a good custom when they see one—and surely celebrating another New Year can be fun.

Facilities and Access

Hours: Although some restaurants are closed on Sunday, the weekend mornings and afternoons seem the best and most enjoyable time to view the neighborhood scene.

Transportation and Parking: The Argyle el stop is right in the middle of "Little Saigon." There is also an exit off Lake Shore Drive for Lawrence Avenue, which is one block south of Argyle Street. Metered parking is difficult on Argyle. The side streets are much easier to park on, but be sure to lock up the car since theft is not uncommon in this neighborhood.

Information: For information about events in the Argyle Street area, contact the Vietnamese Association,

4833 North Broadway, Chicago, Illinois 60640, (312) 728-3700. The association sponsors the Tet New Year's celebration and the Mid-Autumn Children's Festival, which is held in September and includes a dragon dance, outdoor lantern procession by children, and dance and music performances by children. Free candy is often distributed.

The Chinese Mutual Aid Association sponsors the Argyle Street Festival, usually scheduled in August. This major festival includes a lion dance, games, and outdoor food concessions. Contact the association at 1000 West Argyle, Chicago, Illinois 60640, (312) 784-2900.

CHINATOWN

Wentworth Avenue between Cermak and 24th Streets on Chicago's near south side may not be China, but it certainly doesn't seem like Chicago. A weekend morning in Chinatown, when all the Chinatown residents are out shopping and enjoying their home neighborhood, is the best time to experience the area. Chinatown has such an incredible liveliness of colors and sounds that time spent here can be quite exciting for children: Chinese street signs, Chinese shop signs, Chinese architecture, Chinese food, Chinese language, Chinese people—all in Chicago—all in Chinatown.

Bakeries and Gift Shops

It's not too surprising that the gift shops and bakeries are favorite places for children. Not only do these places have the toys and sweets, but they also have Chinese children! For fine baked goods try Happy Gardens, featuring a wide selection of traditional rolls and cookies, 2358 South Wentworth Avenue, (312) 225-2730, and 227 West Cermak Road, (312) 842-7556. The Chiu Quon Bakery—a family-run business—has fun pastries shaped like ice-cream cones, hamburgers, fantastic almond cookies, Chinese soda drinks, and a large sitting area to snack and relax in (2229 South Wentworth, [312] 225-6608).

The Palace Court Art and Gift Shop (2317 South Wentworth, [312] 225-1115) always has children around, and there are plenty of intriguing Oriental toys. Most children enjoy striking the majestic gong situated in the front of the store and investigating the colorful snakes, Chinese yo-yos, flutes, and miniature accordions. Then, if you would like to visit a gift shop geared more to adults, go to Chue Bakery and Gift Shop. They have woks in every available size, china, straw hats, kites, Chinese fans, and even a festival dragon costume for that unusual Halloween party costume. In addition, they have imported canned foods, teas, and various Chinese candies not readily available elsewhere.

Grocery Stores and Restaurants

Several grocery stores in Chinatown are worth a quick visit. Mei Wah Company, 2401 South Wentworth, (312) 225-9090, near 25th Street and on the east side of Chinatown, is a perfect starting point. The shelves are lined with such delicacies as duck sauce, dried seaweed, and ginseng. Unusual smells fill the store and you can watch fresh fish being prepared for sale.

Time to eat, but where should you go? There are numerous restaurants in Chinatown, all fine establishments. Three Happiness at 209 West Cermak, (312) 842-1964, is recommended constantly and for good reason: it mixes good food, friendly waiters willing to answer your questions, and reasonable prices. Another option is the Oriental Palace, 216 West Cermak, (312) 842-5958. Try the dim sum—a selection of Chinese tidbits served like Spanish tapas—to eat until you're full.

For the most authentic Chinese food in the area, try Tin-Yen, 2242 South Wentworth, (312) 842-7156, a popular spot among the Chinese in the neighborhood. While drinking your complimentary oolong tea, peruse the extensive menu and consider following the Chinese custom of ordering one dish for each member of the party and sharing.

You may want to order one of the family-style dinner specials, which will give the children a chance to try several different types of food.

Facilities and Access

Transportation and Parking: Although Chinatown is a safe and pleasant place, the areas directly east and west of Chinatown are dangerous neighborhoods. Car travel is recommended. Take Lake Shore Drive to Cermak Road and travel west for about a mile. There are several parking lots off Cermak, and usually there is parking in the neighborhood directly west of Wentworth Avenue.

GREEKTOWN AND LITTLE ITALY

The old near west side, home of the Greek, Italian, and Jewish ghettos, has undergone a dramatic transformation in the last decades. The construction of the Eisenhower Expressway, the University of Illinois at Chicago, and the general upward mobility of these ethnic groups have made the names Greektown and Little Italy relics of the ethnic past. What survives today of the neighborhoods are restaurants, shops, and some "difficult to budge" older Italian families who still sit out on their front stoops. However, if you and your children are famished and in search of some ethnic flavoring, don't pass up the fantastic Greek and Italian spots on the near west side.

Greek Restaurants

Frequently, Greek and Latin language classes in Chicago make a field trip to one of the Greek restaurants that dot Halsted Street from Van Buren to Monroe. These restaurants provide a great cultural and culinary experience for students. Two restaurants, in particular, are a cut above the rest—The Greek Islands and The Parthenon.

The Greek Islands (200 South Halsted, [312] 454-1660) is a crowded, noisy, fun place that tries hard to re-create the flavor of its native land. Children will be struck by the unusual atmosphere, the "strange" language spoken by the Greeks in the restaurant, and the main, eye-catching, tourist attraction—the flaming saganaki. How daring are you and your children with food? Besides the firey saganaki appetizer—flamed fried cheese—you should consider the chewy octopus, the scrumptious fried squid, and the fresh fish roe. If those selections are unpalatable to your children, there are several more familiar dishes: gyros, shish kebab, shrimp plates, and salads. The fantastic homemade baklava and creme caramelle desserts or Greek cakes are generally pleasing to even the most picky eater.

The Parthenon (314 South Halsted, [312] 726-2407) is a more subdued but carefully authentic restaurant. The moussaka, pastitsio, and dolmades are all excellent and moderately priced, and the family-style dinner offers a variety of Greek foods to pass, share, and taste. We were told the lamb's head is very good and is quite authentic, but for visual reasons you might want to order something a bit more appetizing. The galaktobouriko, Greek crispy phyllo dough filled with custard and baked in syrup, is superb. Several other authentic Greek restaurants feature hearty soups and more elaborate service—including belly dancers.

Greek Grocery Stores and Bakeries

The Athens Grocery (324 South Halsted, (312) 454-0940) provides an opportunity to pick up some of your own ingredients for a home-cooked Greek meal. Kids will enjoy listening to the Greek music; smelling the powerful new aromas; and seeing the mounds of halvah, smoked fish, tubs of cheeses, and numerous types of olives not usually carried by our antiseptic supermarkets.

The nearby Pan Hellenic Bakery, 322 S. Halsted, (312) 454-1886, features a sparkling display of Greek and French pastries and cookies quite unlike any others in town. The Athenian Candle Company, too, is an interesting shop (300 South Halsted, [312] 332-6988). The air is filled with a heavy incense mixture and you can shop for beads, soaps, lotions, and potions.

Italian Restaurants and Shops

Food stores in the old, overcrowded Italian ghetto used to overwhelm the streets with the smell of garlic, sausages, and cheeses. Now, the area has been trimmed down by the construction of the University of Illinois campus, but there are enough great restaurants, bakeries, and Italian atmosphere here to make a trip to Taylor Street an outing. Kids usually love Italian food (if not to eat, at least to play with), and walking down Taylor Street, west of the university, gives children a feeling of what an old ethnic neighborhood was like.

There are a number of good restaurants if you're considering only the quality of food. However, several places offer great food with gruff service. When you are with your children, the atmosphere of hospitality and a patient staff are equally as important as good, healthy food. Florence Restaurant (1030 West Taylor Street, [312] 829-1857) and Febo's (2501 South Western Avenue, [312] 523-0839) specialize in such friendly, pleasant service with good food, too.

Florence Restaurant, opened by Florence Scala and her brother, Mario Giovangelo, is beautifully decorated with antique woodwork, a high tin ceiling, and a lovely contemporary mural. All the pastas are wonderfully cooked and the special desserts will delight children. Febo's offers a lengthier menu than Florence's but has a less charming atmosphere. The veal dishes and tortellini are excellent.

Mategrano's (1321 West Taylor Street, [312] 243-8441) and Gennaro's (1352 West Taylor Street, [312] 243-1035) are decent alternatives, and Mategrano's offers a kid-pleasing buffet on Saturday with more than a dozen items.

While touring the neighborhood on a hot day, the children will be magnetically drawn to Mario's Italian Lemonade (1070 West Taylor Street). Mario's offers scrumptious, iced Italian lemonade in 13 sizes of paper cups and more than a dozen flavors not typical at more standard ice cream shops, including peach, watermelon, and cantaloupe.

Another spot that magically draws children is Original Ferrara, Inc. Bakery (2210 West Taylor, [312] 666-2200). This enormous bakery has the largest selection of Italian cookies, pastries, and cakes in the area. The cannoli—sweet ricotta cheese with pistachios in a crisp dough—is a definite must.

Before heading home, consider a simple stroll around the neighborhood (be cautious after dark). Strike up a conversation with some of the older Italian folks hanging out on their stoops and ask them about the "old neighborhood" and how things used to be. If your children are older, they may be fascinated with the changes, decline, and redevelopment of the city neighborhoods.

Facilities and Access

Transportation and Parking: Drive to this area via the Eisenhower Expressway exiting at the University of Illinois Medical Center. Or head west from the Loop on Harrison, turn south on Halsted to Taylor Street (1000 south). The CTA can give you directions on public transportation (37 Sedgwick bus). Because of convenience and safety considerations, we recommend driving. Parking is available in numerous guarded lots and on the street.

INDIA ON DEVON

If India was the jewel in the crown of the British Empire then the Indian community between Maplewood and Washtenaw is the jewel of Devon Avenue—or maybe of this ethnic city. This area begins about three blocks west of Western Avenue. Curry is what you smell, music of the sitar is what you hear, and women clothed in traditional saris are what you see. Within the perimeter of a few Chicago blocks, you and your children can see the native sounds and smells and clothes of a culture thousands of miles away. Your children can observe the foreign language, dress, and music, while passing the familiar, yet distinctly Indian, video stores, markets, and fast-food restaurants.

Restaurants

Most of the neighborhood restaurants serve buffets, ranging from $4.75 to $5.50 per person (usually half-price for children under six). The absence of beef and dominance of vegetarian fare provide a good vehicle for explaining some concepts of Hinduism. Staff people seem willing to take time to explain what foods they serve and what foods children may like the best and to answer questions about their culture. At some restaurants you can watch food being prepared.

The Gandhi India (2601 West Devon, [312] 761-8714) and Standard India (2546 West Devon, [312] 274-4175) restaurants' buffets are quite similar, including: homemade naan, a bread similar to pita; curried chicken and lamb; basmati rice; hot and cold vegetable dishes like alu gobhi, made with potatoes, peas, and cauliflower; chana masala, made with garbanzos, tomatoes, and spices; and a wide variety of delicious sauces and dressings to sample and mix. If the spices get a little too hot, there are salads of tomatoes, cucumbers, and onions with a cooling cucumber-and-yogurt dressing. Desserts consist of different-flavored dumplings made with cottage cheese and sweetened milk. Some are deep-fried, and all are differently flavored, such as the rasmalai flavored with pistachios, peanuts, and almonds in a cream sauce, or the gulab jamun flavored with honey. The two musts are the mango shakes at the Gandhi India Restaurant and the lassi, a yogurt drink, available at the Standard India Restaurant

The buffet style gives you and your children a good sampling of Indian cuisine. But the real flavor and taste of Indian culture comes from the Indian music and the informative staff.

A number of takeout restaurants also specialize in Indian

and Pakistani cuisine, so you can take the food of this neighborhood into your own.

Stores and Markets

If you and your children are ready to walk off a delicious meal or simply stroll, be sure to stop at one of the stores selling saris. There are rows and rows of beautiful fabrics to see and touch. The hanging displays catch the light from the window to attract your attention as well as your child's. In the stores you can watch sari-clad customers argue over the merits of new fabrics and designs. It is also quiet enough in these shops to really hear the language.

The neighborhood grocery stores are an adventure in themselves. The Jai Hind (2658 West Devon Avenue, [312] 973-3400) and Patel Brothers markets (in the 2500 and 2600 blocks of West Devon) have rows filled floor to ceiling with grains, vegetables, mangoes, spices, sauces, oils, and soaps—all a bit exotic to the Western table. The aroma of curry and the sounds of Indian music permeate the backgrounds of these shops. The shops seem more open and sunlit than most stores our children are accustomed to and transmit the feeling of open-air markets. The shelves are clearly labeled, so if your child is learning to read phonetically, there are new sounds to practice: pooha, chori, kalachana, and kabulichana. Here, too, the salespeople seem very willing to show you around and answer your questions.

Facilities and Access

Hours: Many of the stores are closed on Tuesdays, so you may want to call before taking a trip to this neighborhood.

Transportation and Parking: Places to park are scarce but available on Devon; your chances are better on the many side streets with metered parking. It's best to drive to this area, although the 155 Devon bus (which you can pick up at the Loyola el stop) heads through this area.

Information: For more information regarding hours, buffets, and children's prices, call Standard India Restaurant, (312) 274-4175, and Gandhi India Restaurant, (312) 761-8714.

MEXICAN INTERLUDE—PILSEN

It's not Mexico, but the smells, sights, and sounds will tell you otherwise. Just southwest of the Loop lies the colorful neighborhood of Pilsen. Bounded by the Burlington Railroad tracks, the Chicago River, and Marshall Boulevard, Pilsen is home to Chicago's vast Mexican community. Eighteenth Street, Ashland Avenue, and Blue Island are its main commercial streets. Although the area may appear a bit rough, it's quite safe for an afternoon outing. With children, the best bet is to spend the day at the Mexican Fine Arts Museum (see the Ethnic Museums section) and the beautiful Harrison Park next to the museum. Afterward, venture down the commercial boulevards and check out the bakeries *(panaderias)*, supermarkets *(supermercados)*, stores *(tiendas)*, and restaurants *(y restaurantes)*.

Restaurants

The neighborhood is filled with taco stands *(taquerias)* that provide good, cheap bites to eat. But if you have a little time and a good appetite, head to Nuevo Leon, 1515 West 18th Street, (312) 421-1517. The menu offers the traditional tacos, flautas, and enchilladas along with some Mexican delicacies like goat-, liver-, and tongue-filled tacos. The seating area is clean, intimate, and packed with Hispanic adults and children. The friendly waitresses allow the children to play a bit in the restaurant and the sound of the children combined with the salsa music on the jukebox creates quite a stir in the restaurant. It's worth the noise, however, considering the quality and quantity of food Nuevo Leon offers. It would be difficult to find better Hispanic cuisine in the city.

If Nuevo Leon is filled, consider going down the street to Cuernavaca, at 1158 West 18th Street, (312) 829-1147. This restaurant also has fine food at good prices, but it is more Americanized than Nuevo Leon.

Bakeries

There are several bakeries in Pilsen. They all feature the characteristic unfilled sweet bread topped with sugar or glazes (for a mere 20 cents). The kids will enjoy the butterfly cookies, the various almond cookies, and the different doughnuts offered in the bakeries. The best of the lot is probably Nuevo Leon Bakery (same name as the restaurant) located at 1634 West 18th Street, (312) 243-5977. The method of selecting your own goodies with tongs and a tray is typically Mexican—and fun!

Facilities and Access

Transportation and Parking: The best way to get to Pilsen is by car. Go to Roosevelt (1200 south) and Michigan Avenue and go west on Roosevelt until Halsted Street. Then travel a half mile south on Halsted to 18th Street. Metered parking on 18th Street is difficult so you might have to try the side streets for spaces.

7

PLAZAS, OPEN SPACES, QUIET RETREATS, AND HIGH PLACES

Sometimes museums are just too confining for children, and little ones need a place to expend pent-up energy or relax. Chicago is filled with wonderful and secluded plazas and outdoor retreats. A fun activity for a clear, sunny day is to go to the top of one of Chicago's famous skyscrapers. Or, various gardens and fountains around the city offer quiet places to rest in scenic settings.

During the business week, many plazas have midday programs of music, dance, arts and crafts shows, farmers markets, and fashion shows. The *Chicago Tribune* and *Chicago Sun-Times* carry special columns on plaza events during the summer, and the city maintains a telephone service: dial (312) 346-3278 (F-I-N-E-A-R-T) for information. If you want to avoid the crowds, we recommend weekend visits. But during the week, mid-mornings and mid-afternoons are less crowded than lunchtimes.

The less-crowded plazas lend themselves to all sorts of outdoor activities, including picnicking, skateboarding, roller skating, and kite flying. Chicago has its own kite-flying club—the Chicagoland Sky Liners, (312) 735-1353. They give classes in kite making and kite flying and schedule flying meets in the parks during the summer.

AMOCO OIL BUILDING
200 East Randolph Street
Chicago, Illinois 60601

Although this building fronts Randolph Street, it is interesting to approach it from Michigan Avenue by way of Lake Street. Walk eastward on Lake Street—this is a hearty uphill walk. At the end of Lake Street turn right and walk south along the Amoco Oil Building: it seems as if there is no entrance, but about halfway down the block on your left, steps rise to one of Chicago's most beautiful and secluded plazas. You can enter the building from almost any direction at plaza level. It really deserves a full walk around to appreciate the careful attention to landscaping. The waterfalls with seats around the edge, the flowers, and the many shade trees are all well secluded from the noise of the city. From the plaza level, you can enter through revolving doors and take the escalator to the many shops, restaurants, a branch of the U.S. Post Office, and restrooms.

From the lower level, come to the south end of the building and exit into another serene and beautiful plaza. This small but shaded plaza with trees and flower beds has three waterfalls pouring down the expanse of the south wall—you are surrounded by the sound of rushing water. In the middle of this plaza is a large reflecting pool with a unique sculpture, *Whispering Rods,* by Harry Bertoia. If you listen carefully, you'll discover that the rods really do whisper as the constant breeze causes them to collide and rub against each other in a musical chime. This plaza is virtually empty on weekends and is perfect for a picnic, roller-skating, or just reading a story. Look up and see the cityscape as it marches southward on Michigan Avenue.

ART INSTITUTE GARDENS
South Michigan Avenue and East Adams Street Chicago, Illinois 60603

The Art Institute boasts five fountains and three lovely gardens, each plaza-like. The quiet garden on the south side has many shade trees and benches. There is a reflecting pool in front of a Lorado Taft sculpture, *Spirit of the Great Lakes,* in which figures holding a giant shell represent each of the Great Lakes. The sculpture originally graced the south wall of the institute. The ground is gravel-lined, so it is not conducive to running.

At the back of the Art Institute, along Columbus Drive, the famous arch from the old Board of Trade Building serves as the entrance to a lovely picnic area, with fountains and terraced grounds. The picnic area is small but shaded and gravel-lined. A bit further south is yet another large reflecting pool with a fountain—this one titled "Celebration," designed by Isamu Noguchi. Enter the Art Institute from here for ready access to the three restaurants in the building. In summer, a special visit to the quiet Garden Restaurant, with its center pool, sculpture, and umbrella tables, is a special treat. There are lockers here to store your parcels, and you can procure a stroller if you wish to view the museum.

There is an elevator to the second-floor dining room, which is the most expensive restaurant in the building. The lower level contains the cafeteria, the Garden Restaurant, and restrooms.

BUCKINGHAM MEMORIAL FOUNTAIN AND PLAZA
(At Congress Expressway and Columbus Drive)

Walk eastward up Congress from Michigan Avenue past the famous *Spearman* and *Bowman* statues to the grandest of all plazas in Chicago. If you go on the hour you will be rewarded by the spectacular water display—at night the show is further enhanced with colored lights. From the plaza you can view the harbor and the bobbing sailboats. Walk beyond the fountain on all four sides to view the four companion fountains: Crane Girl, Dove Girl, Fisher Boy, and Turtle Boy. Here also are several new sculptures and the famous rose gardens. During the week there is usually a hot-dog vendor or two on the plaza, and the whole area is the site of many summertime activities.

DALEY BICENTENNIAL PLAZA
(South from the Amoco Oil Building to Monroe Street)

(Note: don't confuse this with the Daley Center Plaza on Washington Street.) Walk eastward along Randolph Street from Michigan Avenue for a truly magical view of the city. You will see the Art Institute to the south and the arch (from the Old Chicago Board of Trade Building) that now graces one of the plazas behind the Art Institute. From this vantage point you can see the Chicago skyline being developed along the lake, the luxury condos of Lake Point Towers to the north, and Buckingham Plaza and Outer Drive East directly ahead. If you cross over Columbus Drive, you'll arrive at a parkway with an entrance to the Monroe Underground Parking—turn to your right and directly before you at the far end is the magnificent Buckingham Fountain, in all its glory every hour on the hour in the summertime. Here you can view the two unique wildflower ovals that were planted for the city. Volunteers have meticulously identified and labeled most of the flowers, which are native to Illinois and were probably in a meadow during the time of Fort Dearborn. It is a breathtaking sight when the wildflower gardens are in bloom, from early summer well into the fall.

This park is named the Richard J. Daley Bicentennial Plaza in honor of the late mayor. In the sky to your left you will note the many small airplanes approaching and departing Meigs Field. The wall here is flanked on either side by stairs leading down to the parkway—you'll see picnic tables and many benches. This area is also easily accessible by bicycles and strollers. There are public tennis courts on either side. From here you can walk past the Petrillo Music Shell and the famous rose gardens that skirt the fountain to Buckingham Fountain and beyond. It is a magnificent and panoramic view of Chicago's ever-changing skyline and one of our personal favorite views.

DALEY CENTER PLAZA

(At Washington, Dearborn, Randolph, and Clark)
55 West Washington Street
Chicago, Illinois 60602

This is, perhaps, Chicago's most noted plaza. Here in silent majesty rises the famous but untitled Picasso sculpture. Glance across the street (Washington) to a smaller plaza to see another famous lady—Miro's *Chicago*. The plaza fronts the city and county courts as well as the Cook County Law Library. It is popular all year-round with numerous activities, including lunchtime programs and Chicago's official Christmas tree lighting, usually on the Friday after Thanksgiving. A great free view—with no wait—of the city greets visitors willing to go through the airport-like security check and peek out the 26th-floor windows of the Daley Center.

The single tree and benches on the south corner are largely inhabited by pigeons and are usually in rather filthy condition, but the benches near the fountain and pool are usually clean and more inviting. Children love to climb and slide on the sculpture in this large, flat plaza.

At the Washington Street end is the eternal flame dedicated to the men and women of the armed services. A very busy plaza during the week with lawyers crisscrossing to City Hall, the courts, and the county buildings, it is almost deserted on the weekends. If you take the escalators inside the building you can go underground to City Hall and connect to Chicago's lengthy underground pedway system, on which you can travel all the way to Michigan Avenue. Eventually all major buildings will connect to this underground system. Underground maps are available at City Hall.

FIRST NATIONAL BANK PLAZA
(At Dearborn, Madison, Clark, and Monroe)
One First National Plaza
Chicago, Illinois 60603

The First National Bank Plaza is well terraced and tree-lined. In summer it has a small outdoor cafe on the Dearborn side. From street level take the steps down to the first level to view Chagall's beautiful mosaic, *The Four Seasons,* of hand-chipped stone and glass fragments. Benches surround the monument. Proceed further down to the lower level to the gazebo stage, which is host to many lunchtime concerts during the summer. A magnificent fountain spouts from the floor of this plaza. Look straight up and you will have a unique view of Chicago's varied downtown architecture, including the First National Bank Building itself. This secluded, below-street-level plaza imparts a quiet serenity. Note also the clock that dominates the south end of the second level—a favorite lunchtime meeting place.

FOURTH PRESBYTERIAN CHURCH GARDEN
126 East Chestnut Street
Chicago, Illinois 60611

When you are visiting the Water Tower Place or the John Hancock Building, you may wish to cross the street and walk through the quiet garden of the Fourth Presbyterian Church. It's beautiful at Christmas time, when the garden arches are decorated with Italian lights. In summer, the garden's small, melodious fountain is the scene of many weddings. In this little-known retreat from the noise and clamor of Michigan Avenue, your children can roam within the safe confines of the garden while you recuperate for the rest of your day.

HEALD SQUARE
(At State Street and Wabash Avenue)

This is not really a square but is the intersection of Wacker Drive, State Street, and Wabash Avenue. It angles between State and Wabash. At the State Street corner there is a low bubbling fountain that is dedicated "to the men and women of Chicago who served their country with honor and distinction in the Vietnam War." The steps behind the fountain lead into a tiny park lined with blue-spruce evergreens, flowers, and stone benches. Unfortunately, this is also one of the city's most unkempt plazas. Past the park you will find a drinking fountain.

At the Wabash end is the Robert Morris, George Washington, Haym Salomon monument designed by Chicago's Lorado Taft. It was presented in 1941 by the Patriotic Foundation of Chicago and it was dedicated on the 150th anniversary of the ratification of the American Bill of Rights. Morris and Salomon, the civilian patriots, stand hand in hand with Washington. Each gave financial support and made personal sacrifices to assure victory to the United States in the Revolutionary War. This memorial symbolizes the historic fact that people of different faiths and origins worked together and fought side by side to create our nation. This square is a Chicago landmark.

From this wonderful plaza you have a superb panoramic view of Chicago's Michigan Avenue buildings: the Wrigley Building; Tribune Tower; Marina City; the unusual white Seventh Church of Christ, Scientist; and to the west, the Merchandise Mart. Note particularly the ornate clock attached to the building to the south of the square.

The easternmost part of Heald Square is home to one of Chicago's most enchanting fountains—the Children's Foundation with its filigreed and sculpted four tiers and gargoyle heads, cherubs, children, cranes, and seashells. Designed by Burnham, it was finally installed by Mayor

Jane Byrne in 1982. It's always a delight to see children splashing in the fountain, and remember to bring dry clothes in case your kids want to join in the fun.

Across the street is the Chicago River and marina at Marina Towers. A block farther, at the corner of Wacker and Michigan, one can descend steps on either side of the bridge to the docks of the Wendella and Mercury boat cruises that go out on the lake. The double-decker tour buses stop here every 15 minutes to pick up and drop off people at the boat rides. At either end of the Michigan Avenue bridge, huge pylons with their elaborate bas-reliefs are dedicated to Chicago's discoverers and pioneers.

HUTCHINSON FIELD

(At Columbus Drive and 9th Street)

After touring the Field Museum or the Shedd Aquarium, this is a most welcome spot to kick off your shoes and walk through the grass or just visit on the weekend. Just to the north of the Field Museum, Hutchinson Field is unique because it is sunken with raised edges all around. Children can roll down the sides and run to their hearts' content. The wind currents make this large field a super place to fly kites. Two sides are bordered with magnificent cherry trees that blossom in profusion in early spring. This field, for some reason, is little-used—which makes it a perfect spot to have to yourself. From here you can view the lake, the museum to the south, and the skyline of Chicago to the north. Bring food and drink if you plan to stay, since the field is quite a distance from any amenities.

JOHN HANCOCK BUILDING

875 North Michigan Avenue
Chicago, Illinois 60611
(312) 751-3681

The John Hancock Building, Chicago's second-tallest building, boasts a skydeck from which you can enjoy a breathtaking view of the lake and Chicago at large. The view is especially rewarding at night, when all the lights trace the city's many streets and thoroughfares. Hours: 9:00 A.M.–12:00 midnight, every day. Admission: adults, $3.50; children ages 5–15, $2.00; children under 5, free.

Also, consider going to the lounge on the 96th floor. While it is not an environment intended for children, it is very pleasant in the early evenings. Families can drink sodas and admire the view for about the same price as admission to the observation deck. The lower levels of the Hancock building abound with small shops and restaurants.

MCI PLAZA

(At Michigan Avenue and Lake Street)

At the corner of Michigan and Lake you will see MCI Plaza that skirts Boulevard Towers, which is part of the Illinois Center complex. It is another beautifully landscaped, small plaza with flowers, shrubs, trees, and sitting benches. The large, multicolored outdoor sculpture titled *Splash,* by Jerry Peart, was erected in 1986. Children will delight in its bold colors and shapes. Enter the revolving doors of Boulevard Towers to explore the many shops and restaurants of Illinois Center—you can travel underground from Michigan Avenue, through Illinois Center and the Hyatt Hotel to Columbus Drive, and 3 Illinois Center.

PIONEER COURT

(At the Equitable Building)
401 North Michigan Avenue
Chicago, Illinois 60611

The base of the fountain in this court is inscribed with the names of Chicago's pioneer families and founders—Jean Baptiste DuSable, Charles Henry Wacker, Montgomery Ward, Jane Addams, Philip Armour, and Daniel Burnham. The names are beginning to fade a bit, but the fountain deserves a walk around just to discover the names of those who had a vital part in the building of Chicago. The fountain is wonderful for sitting at the edge and cooling one's feet in the frothy waters. Numerous circular stone seats surround the enormous flower holders and trees, and many musical events occur here during the week, sponsored largely by the *Chicago Tribune.*

Pioneer Court is a very spacious, flat plaza—great for bikes and kites. It is one of the few plazas laid with brick, so it is not recommended for roller skating or skateboarding. Walk to the back of the plaza for a view of the new construction going on around North Pier. At the south end one can view the Chicago River and, beyond the river, the construction along East Wacker Drive—the Hyatt Hotel, the new Swiss Hotel, and the Illinois Center complex. Here you can watch yachts, cruise boats, barges, and occasionally, a crew team. A special reward in early spring is the not-uncommon sight of mother wood ducks with their baby ducklings out for a river swim.

It was here, at the mouth of the Chicago River, that Père Jacques Marquette and two voyageurs built a shelter to camp. This plaza was the original site of what is now Chicago.

The low glass building in front of the Equitable Building houses an escalator system that goes underneath the building to shops and restaurants. This is an opportunity to introduce children to revolving doors and escalator systems when there aren't a lot of people around: early morning or mid-afternoon is best, because the glass building is generally closed on weekends.

At the south end of this plaza, just before crossing the bridge, is a stairway descending to the river level, where there is a delightful outdoor restaurant. The steps leading down are very small and easily negotiable by small children, and descent with a stroller is manageable.

If you walk south across the bridge on the easternmost walkway you will discover, at the corner of Michigan and Wacker, bronze plaques embedded in the sidewalk that outline the original site of Fort Dearborn.

If you walk north on Michigan Avenue, you can read off

the signs on the parts of history embedded in the Tribune Tower—stones from famous structures worldwide.

SEARS TOWER PLAZA AND SKYDECK

233 South Wacker Drive
Chicago, Illinois 60606
(312) 875-9696

Enter the Sears Tower from either Wacker Drive or Franklin Street. The Wacker entrance is more impressive, for it passes through the massive atrium with twin circular stairs leading up to the huge Alexander Calder mobile, *Universe,* that has five moving parts. Two waterfalls run beneath the staircases, and steps lead down to the many shops and restaurants that populate the Sears Tower. Information booths are at both entrances. Be sure to ask for a brochure on the Sears Tower: the statistics quoted inside the brochure are quite impressive.

The skydeck is best reached from Franklin Street. You can either take the stairs down to the ticket booths or ride the elevator if you have strollers or wheelchairs. Admission: $3.75 for adults ages 16 and over; $2.25 for children ages 5–15; and no charge for children ages 4 and under. A special family pass is $9.00. Hours: 9:00 A.M. to 12:00 midnight, every day.

On mid-summer afternoons, long lines for the skydeck and restrooms are common. The restrooms are quite small and have no amenities for babies or small children—not even counter space for changing diapers. Restrooms are not listed on the several directories, so you will need to ask directions from the information booth attendants to find them. If you see tour buses lined up outside on Wacker Drive, you can be sure of long lines and generally crowded conditions!

On the south side of the Sears Tower is a very large, open plaza, largely undecorated except for the huge flower containers and curved benches. There is no shade here, but the plaza is interesting because it slopes at a 45° angle from Wacker Drive east to Franklin Street and would probably be a challenge to the practiced skateboarder or roller skater. During the week the plaza is crowded for the lunch-hour music and other activities. Empty on weekends, the plaza always has a breeze sweeping through it and a sense of privacy you might enjoy.

At the high end of the plaza (Franklin Street) one can view several new buildings under construction, the old Post Office building, and to the far south, River City. Directly across the street at the self-park garage is a cluster of the famous red, double-decker tour buses that provide one-hour tours of the city ($5.00 for adults; $2.00 for children under 12). This is the main boarding place for the city tour.

STATE OF ILLINOIS BUILDING PLAZA

(At Clark, Randolph, LaSalle, and Lake)
160 North LaSalle Street
Chicago, Illinois 60601

At the north end of the Daley Plaza you will see the glass protuberance of the State of Illinois Building, designed by Helmut Jahn. In front is the black and white sculpture, *Monument With Standing Beast* by Jean Dubuffet, erected in 1984. Children love to play in this sculpture, and it is the only sculpture in the Loop seemingly created just for them. The outside plaza is small, but the real expanse is within the building itself. Enter through the revolving doors behind the sculpture to behold the immense rotunda with glass elevators mounting skyward.

Take the escalator down to many shops, fast-food restaurants, and spacious restrooms. Note especially the waterfall underneath the escalators. The building is not open on weekends, however. A ride up these glass elevators for the high view is well worth the trip.

U.S. POST OFFICE PLAZA
(At Jackson, Dearborn, Adams, and Clark)

Located in the center of downtown, this plaza is one of the two frequented by pigeon-feeders who, fortunately, confine their activities to the north end of the plaza. The huge Calder sculpture, *Flamingo,* gracefully arcs over the plaza, and several benches are available throughout the area. A model and historic information about building the plaza are located across the street in the Dirksen Federal Building.

This is another windy plaza, where children may thrill to the many whirlwinds that constantly play under the sculpture and provide opportunities to explore wind currents. At the southwest corner (Jackson and Clark) another sculpture, *Ruins III,* silently guards a small garden surrounded by benches and shade trees. Deserted on the weekends, this huge, flat plaza is good for many activities.

There are a number of parking garages south of Jackson. To the north, in the basement of the LaSalle Bank Building, is one of the few remaining public cafeterias in the Loop. Inexpensive, it features a salad bar, taco bar, entrees, pastas, sandwiches, and carry-outs. There is also a small diner on the street level on the LaSalle side.

This is the heart of Chicago's financial district, with the Continental, LaSalle, and Harris banks and the Bankers Building lining Clark Street.

WATER TOWER PLACE
835 North Michigan Avenue
Chicago, Illinois 60611

This is not really a plaza, but children will probably delight in the escalators, revolving doors, and glass elevators that silently glide through the center of this huge shopping area. They will especially like FAO Schwartz, the toy store, with the huge, furry, stuffed animals and shelves overflowing with every toy new to the market. Water Tower Place also has many restaurants (see directories on each floor), restrooms, souvenir shops, and an underground parking garage, as well as several movie theaters.

Water Tower Place is not conducive to strollers, because the elevators are small and slow in coming. The building is crowded year-round—especially at the holidays—with shoppers and tourists.

8
BEHIND THE SCENES AT CHICAGO'S INSTITUTIONS

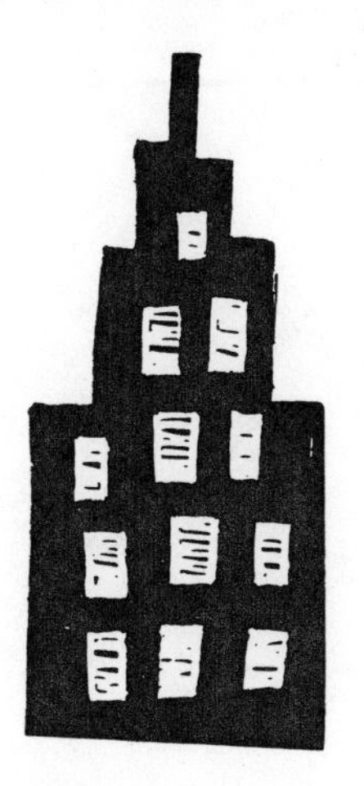

BEHIND THE SCENES AT THE CHICAGO SYMPHONY ORCHESTRA

220 South Michigan Avenue
Chicago, Illinois 60604
(312) 435-6666

To facilitate a child's fascination with classical music of the Western World, a visit to Orchestra Hall may be worth a trip.

Behind-the-scenes tours are usually scheduled on three dates in the fall at a cost of approximately $5 per person. This historic building has hosted the greatest artists in symphonic music since 1904 and reeks of a musical atmosphere.

A tour will guide you through Maestro Solti's dressing room backstage, the music library, and the broadcast and recording facilities, which are actually used to record live performances. The tour often includes a ride on the stage's hydraulic piano lift, which swishes mammoth grand pianos on and off the performing space with grace. Refreshments are included.

If you subscribe to the symphony or have friends or relatives who do, they will often receive cards granting access to the Chicago Symphony Orchestra's afternoon rehearsals free of charge. If your children might enjoy this adventure, ask your friends for their admission cards—most working people routinely toss them away. Seeing the musicians laboring to achieve perfection in their performance is, in itself, an experience most children easily identify with and appreciate.

BEHIND THE SCENES AT THE EXCHANGES

Older children and even younger ones in a family connected to the markets may enjoy the bustle and frenetic activity of one of the trading pits in Chicago. Open only during the trading day, each of the following provide a half-hour amusement and are in the heart of the financial district.

Chicago Board of Trade

141 West Jackson Boulevard
Fifth Floor
Chicago, Illinois 60604
(312) 435-3590

The oldest of three major exchanges in the city, established in 1848, the CBOT trades agricultural commodities and financial products in approximately 15 pits. Beginning at 9:00 A.M., and every hour on the hour through 12:00 noon, the CBOT conducts a 30-minute tour consisting of a 15-minute movie and a verbal explanation of what is happening on the floor. Questions are encouraged, but many have no answer, such as, "How do you buy low and sell high?"

Chicago Board of Options Exchange
400 South LaSalle Street
Fourth Floor Viewing Windows
Chicago, Illinois 60605
(312) 786-7492

Trading in options to purchase, the CBOE's glitzy new structure is a self-contained shopping mall along with a viewing gallery where children can observe the grown-ups making and losing untold fortunes. Guided tours are available to high-school groups, and literature is available.

Chicago Mercantile Exchange
30 South Wacker
Fourth Floor Gallery
Chicago, Illinois 60605
(312) 930-8249

A total of 32 commodities are traded at the Merc, including livestock and world currencies. The viewing room is open from 7:30 A.M.–3:15 P.M. daily and contains interactive video machines, which explain what's going on, to play with. Guided tours can be arranged by calling ahead and requesting the service.

BEHIND THE SCENES AT THE LYRIC OPERA OF CHICAGO
20 North Wacker Drive
Chicago, Illinois 60606
(312) 332-2244

Children are naturally attracted to the opera, and the Chicago Lyric Opera offers backstage tours approximately six times a year between November and February, during its season. You and your children can tour a real prima donna's dressing room, the costume department, the armory, the catwalks above the sixth floor, the orchestra pit, and the lifts and wheels used to shift scenery and stage heights.

This multimedia art form includes music, dance, costumes, and often humor, but children (and adults) are sometimes put off by the fact that many operas are sung in another language. The Lyric Opera projects the English translation during its regular performances, but for young ones this does little good. If you plan to attend a performance with your children, try playing some operatic music at home to acclimate your children to the type of sounds they can expect at the opera.

BEHIND THE SCENES AT THE NEWSPAPERS

Fascination with the print media can begin early, since sounding out headlines is breakfast sport for young readers. Both major metropolitan daily papers conduct tours of their production facilities. Also, if your children are really interested in journalism, investigate the small suburban papers and the small-town papers in the communities near the city. The characters who write what we read on a daily basis have newsprint ink in their blood and may relish an opportunity to let your children know that reporting is the most exciting job in the world.

Chicago Sun-Times

401 North Wabash Avenue
Chicago, Illinois 60611
(312) 321-2032

The *Chicago Sun-Times* conducts one-hour tours of its main production center, the news room, the composing room, and the press room on Tuesdays, Wednesdays, and Thursdays at 10:30 A.M. most weeks, with the exception of the entire month of July.

The *Sun-Times* requires its tour members to be at least 11 years old, allows only 25 persons per day to join the tour, and requires reservations.

Chicago Tribune
Tours at Freedom Center
777 West Chicago Avenue
Chicago, Illinois 60610
(312) 222-2116

The *Chicago Tribune* (which calls itself the World's Greatest Newspaper and thus named its television station WGN-TV) also conducts tours of its noisy production facility at 777 West Chicago Avenue. Only the editorial and advertising offices remain at the 435 North Michigan Avenue location, which transmits all copy electronically to the press room, circulation, and distribution center on Chicago Avenue.

The tours are scheduled on weekdays at 9:30 A.M., 10:30 A.M., 1:30 P.M., and 3:30 P.M. and run for 45 minutes. The *Tribune* tour desk advises interested individuals to call for a reservation and requests that tour participants be at least 10 years old. Parking is available on site during the tours.

CHICAGO PUBLIC LIBRARY CULTURAL CENTER
76 East Washington Street
Chicago, Illinois 60602
(312) 269-2900

The Chicago Public Library's Cultural Center is an outstanding resource to use for inspiring children to read, investigate, and imagine. Children love to sit at the miniature tables designed especially for them; to say hello to "Banjo," the huge stuffed bear who is seated along with them in a miniature chair; to look into the Storybook House, where characters like Paddington Bear or Boy Blue hide; and to explore the microcomputer, the audiocassette tapes, and yes, even the shelves full of books.

Children's Library

The Thomas Hughes Children's Library, on the fifth floor of the Cultural Center, contains the largest collection of children's books in Chicago. The more than 60,000 volumes include picture books, easy readers, children's classics, contemporary fiction and nonfiction, and books in foreign languages. The children's librarian helps both kids and adults select books that are specifically targeted for a child of a certain age or for an adult who wants to give a good read-aloud.

Kid Paks

If you work downtown, the Children's Library Kid Pak program is ideal for you. Call the Children's Library at (312) 269-2835, and tell the children's librarian your child's age, reading capability, interests, and library card number, and then he or she will put together a pack of books (and some other small goodies) that you pick up at the Children's Library. It's like ordering a pizza but easier, since you don't have to make the selection yourself.

Children may obtain their own library cards as soon as they are able to print their names. This card permits children access to borrowing all circulating library materials including the very popular stories on audiocassettes. Some of the more popular cassettes, like "How To Eat Fried Worms" or Dr. Seuss stories, are difficult to find since they are often out, so it is a good idea to phone first if you're looking for a specific cassette or book.

Special Events

Even if there is a library in your neighborhood where you prefer to obtain books and materials, do read the Cultural Library's calendar of events for children. Besides the well-known music series for adults, the library also has outstanding programs for children held Saturdays during the school year and on Mondays through Saturdays during the summer. Local theater groups in Chicago put on plays, like *The Enormous Turnip, Rapunzel,* and *Are You My Mother?,* to the delight of a crowd of children. Plays and performances are usually held in the spacious Preston-Bradley Hall on the second floor. The Cultural Center also hosts a wide array of magic acts, puppet shows, movies, musical performances, storytelling hours, and mime demonstrations throughout the year.

If the Cultural Center is accessible to you, do not hesitate to take advantage of the wonderful programs they offer. Call (312) 744-6630 for specific information on the current programs.

Facilities and Access

Hours: The Cultural Center is open Monday–Thursday, from 9:00 A.M. to 7:00 P.M.; Friday, from 9:00 A.M. to 6:00 P.M.; and Saturday, from 9:00 A.M. to 5:00 P.M.

Admission: The library is free for children and adults.

Transportation and Parking: The Cultural Center is accessible by either the el (Washington stop) or Lake Shore Drive (exit at Randolph).

Restaurants and Food: While no food is allowed in the building, the Cultural Center is located on Michigan Avenue, and dozens of pizza parlors, delicatessen-type restaurants, and outdoor cafes are within walking distance.

Restrooms: There are restrooms on most floors, and the fifth floor restrooms (just outside the Children's Library) have changing tables.

Information: For information on programs and exhibits at the Cultural Center, call (312) 744-6630. To reach the Thomas Hughes Children's Library, call (312) 269-2835. The general information number is (312) 269-2900.

HISTORICAL HOUSES

Many of the suburbs have historical houses that re-create the era in which they were built. Some youngsters travel easily through time and can be transported back to that other era by a visit to one of these places.

Your children will probably love to hear about what their days would have been like if they had lived in the houses—how the fires would have had to be lit in winter before they could get out of bed, how they would have dressed for school, how they wouldn't have reached for the box of corn flakes, but would have eaten a different kind of breakfast, how they would have gone to school, what they would have studied, what you as a parent would have done, and how it all differs from what you do today.

Some of the historical homes are beautiful mansions, others are more modest dwellings: each is worth a brief visit for history buffs and their children. Surely, if you must journey to one of the suburbs with your family for another purpose, be sure to inquire if there is a local village or county museum to visit as part of your outing. Listed here are a few that have some special small-scale charm (note: use the 312 area code for all before 11/11/89):

Naper Settlement, 201 West Porter Street, Naperville, Illinois, (708) 420-6010. Open May through October. Wednesday, Saturday, and Sunday, 1:30 P.M.–4:30 P.M. $3.50 adults; $1.75 ages 6–18 and seniors 62 and over; $8.75 family.

Aurora Historical Museum, 305 Cedar, Aurora, Illinois, (708) 897-9029. Wednesday, Saturday, and Sunday, 1:00 P.M.–5:00 P.M. Adults $1.50; children $.75.

Historical Society of Oak Park and River Forest, 217 Home Avenue, Oak Park, Illinois, (708) 848-6755. Saturday and Sunday, 2:00 P.M.–3:30 P.M. Adults $2.00; children $1.00.

Frank Lloyd Wright Home and Studio, 951 Chicago Avenue, Oak Park, Illinois, (708) 848-1500. Weekdays 11:00 A.M.–3:00 P.M.; weekends 11:00 A.M.–4:00 P.M.

Cantigny, One South 151 Winfield Road, Wheaton, Illinois, (708) 668-5161. Free; museum, 10:00 A.M.–4:00 P.M. daily, except Monday. Picnic area and gardens, 7:00 A.M.–8:00 P.M., daily.

MARSHALL FIELD'S

111 North State Street
Chicago, Illinois 60602
(312) 781-1000

Consider an outing to Marshall Field's downtown store. Even if you're an urban resident, your children may be foreigners to the Loop. Traveling downtown by el or commuter train can be a self-contained adventure if it is not routine for your children. And Marshall Field's is a Chicago institution and an archetype of its genre.

Field's Christmas windows are usually decked with music and puppets, and a splendid Christmas tree graces the dining area. The Walnut Room on the seventh floor is a good focal point for planning a pre-holiday visit. But remember what most grandmothers know: circumvent the Christmas crowds by arriving early at the Walnut Room when lunch is beginning, thereby avoiding the wait in line. You can often call ahead for reservations, too, which can eliminate the waiting time for your family.

Most children respond well to the toy displays, the food halls, the decorated windows, the inside store decorations, and the festive feeling of this venerable store during the holidays. Don't even think of serious shopping with your children in tow—just enjoy the ambience of the center of Christmas commercialism in its vertical and urban manifestation.

There is a mother's and children's washroom on the fourth floor, complete with potty seats, a child's-height sink, changing table, crib, and sitting area for nursing that's good to know about if you are in the middle of your day at Field's.

The new and somewhat glitzy lower level of Field's has a food court with four ethnic stations—Mexican, Italian, Chinese, and German—for quick service, which is always necessary with children. As a different sort of more imaginative adventure, my children have always liked the rooms of furniture arranged in the Model Home on the furniture floor. They have created an excellent "pretend" game of what the people who live there after the store closes are really like based on their taste in furnishings. Imagining what these shadow people and their children do during the day (and on Monday and Thursday nights) is another good "pretend" game for your visit to Field's.

If your children like chocolate (and don't get ill from it) consider ending your visit with a purchase of Field's own brand of chocolate truffle, called Frango Mints. Most chocolate lovers born and bred in Chicago use Frangos as a benchmark in the sensuous awakening to the pleasures of chocolates. If you are lucky, and the wind is right, sometimes the fragrance of Bloomer's chocolate factory (located on West Kinzie) may waft through the Loop, further enhancing your exit from this venerable emporium.

Facilities and Access

Hours: Tuesday, Wednesday, Friday, and Saturday, 9:45 A.M.–5:45 P.M. Monday and Thursday, 9:45 A.M.–7:00 P.M. Open one Sunday per month, 10:00 A.M.–5:00 P.M.

Transportation and Parking: Enter from the Howard Street subway station at Washington Street or travel down Michigan Avenue to Randolph Street.

Restaurants and Food: On the lower level: Hinky Dink-Kenna's; and Food Court, with four ethnic bars. On the seventh floor: Walnut Room; It's Italian; English Room; and Bowl & Basket Cafeteria. On the Pedway: The Down Under ice cream parlor is being relocated to the seventh floor and is scheduled to reopen in the summer of 1989.

Restrooms: Each floor has restroom facilities; special mothers' restrooms are on the fourth floor.

Information: Call the store's main number, (312) 781-1000.

UNIVERSITY OF CHICAGO CAMPUS

5801 South Ellis Avenue
Chicago, Illinois 60637
(312) 702-1234

Is it ever too early to introduce children to the concept of college? The University of Chicago will surely have an impact with its stately buildings and quadrangles. Here resides the Oriental Institute Museum, Rockefeller Chapel, the renowned Regenstein Library, and the David and Alfred Smart Gallery. It was at the University of Chicago's Stagg Field in 1942 that Dr. Enrico Fermi and a group of scientists created the first nuclear chain reaction.

David and Alfred Smart Gallery

The Smart Gallery is at 5550 South Greenwood Avenue, Chicago, Illinois 60637, (312) 702-0200. It is closed Monday and open Tuesday, Wednesday, Friday, and Saturday, 10:00 A.M.–4:00 P.M.; Thursday, 10:00 A.M.–7:45 P.M.; Sunday, 12:00 noon–4:00 P.M.. Admission is free.

The Smart Gallery's permanent collection includes Greek and Roman antiquities, and painting and sculpture from the Middle Ages. This is the fine arts museum of the University of Chicago. It was established in 1974 to preserve and exhibit the university's art collection of nearly 7,000 works. The gallery's special collections are the Tarbell Collection of classical ceramics and statuary, the Case collection of early Christian and Byzantine artifacts, and the Epstein Archive Collection of 16th- to 19th-century prints and drawings.

Four major exhibits are presented each year, and tour guides and curriculum aids for primary and secondary students are available. Lecture and film series are also part of the programs.

University of Chicago Concert Series

The University provides a series of musical programs, titled "In Performance," presented by four separate groups. Tickets for the concerts are available at a modest cost from the Department of Music Concert Office, Goodspeed Hall, Room 310, 5845 South Ellis Avenue, (312) 702-8068.

Maps of the entire university may be obtained from the Office of Special Events, Room 200, 5801 South Ellis Avenue.

Just stroll the grounds and be sure to visit Rockefeller Chapel, which is especially festive at Christmas time.

Chamber Music Series: Five concerts per year at Mandel Hall, 57th and University, at 8:00 P.M.

Contemporary Chamber Players: Four concerts per year at Mandel Hall, at 8:00 P.M.

Rockefeller Chapel Series: Three concerts per year including Handel's *Messiah* in December, at 59th Street and Woodlawn. Times vary.

Early Music Series: Three concerts per year at Mandel Hall, at 8:00 P.M.

Facilities and Access

Transportation and Parking: To reach the campus by car from the Loop, exit Lake Shore Drive at 53rd Street. Take 53rd Street west to Ellis, turning south on Ellis. The campus begins around Ellis and 58th Street. By CTA bus, take the Jeffrey bus from the Loop to 55th Street. Transfer to the 55 Garfield bus, getting off at Ellis.

Restaurants and Food: A cafeteria-style food service, with a good selection of healthy snacks and treats, is in Mandel Hall.

Restrooms: Located in Mandel Hall.

Information: The university's main number is (312) 702-1234.